Routledge Reviva

The Land Problem in the
Developed Economy

Land is an important finite commodity in the modern world. In the past wars have been fought over it and land shortage has been the cause of many famines. In modern times debates rage over just how land should be controlled by government and over whether land should be publicly or privately owned.

This book, which was first published in 1984, surveys the major problems and debates connected with land use in the modern developed world. The opening chapters examine the main components of the problem and describe the development of the debate about land from Malthus onwards. The book then analyses land policy in a number of different countries, including the United Kingdom, the United States, Japan, and Eastern Europe.

This book is ideal for students of geography and economics.

The Land Problem in the Developed Economy

Andrew H. Dawson

First published in 1984
by Croom Helm Ltd

This edition first published in 2015 by Routledge
2 Park Square, Milton Park, Abingdon, Oxon, OX14 4RN
and by Routledge
711 Third Avenue, New York, NY 10017

Routledge is an imprint of the Taylor & Francis Group, an informa business

Publisher's Note
The publisher has gone to great lengths to ensure the quality of this reprint but points out that some imperfections in the original copies may be apparent.

Disclaimer
The publisher has made every effort to trace copyright holders and welcomes correspondence from those they have been unable to contact.

A Library of Congress record exists under LC control number: 83024360

ISBN 13: 978-1-138-85319-5 (hbk)
ISBN 13: 978-1-315-72296-2 (ebk)
ISBN 13: 978-1-138-85321-8 (pbk)

THE LAND PROBLEM IN THE DEVELOPED ECONOMY

CROOM HELM SERIES IN GEOGRAPHY AND ENVIRONMENT
Edited by Alan Wilson, Nigel Thrift, Michael Bradford and Edward W. Soja

The Land Problem in the Developed Economy

Andrew H. Dawson

CROOM HELM
London & Sydney

BARNES & NOBLE BOOKS
Totowa, New Jersey

© 1984 A.H. Dawson
Croom Helm Ltd, Provident House, Burrell Row,
Beckenham, Kent BR3 1AT
Croom Helm Australia Pty Ltd, First Floor, 139 King Street,
Sydney, NSW 2001, Australia

British Library Cataloguing in Publication Data

Dawson, A.H.
    The land problem in the developed economy.
    1. Land   2. Natural resources
    I. Title
    333   HC55
    ISBN 0-7099-0819-9

First published in the USA in 1984 by
Barnes & Noble Books, 81 Adams Drive,
Totowa, New Jersey, 07512

Library of Congress Cataloging in Publication Data

Dawson, A.H. (Andrew H.)
    The land problem in the developed economy.

    1. Land use – Government policy. I. Title.
[HD111.D35   1984b]   333.73'13'091722   83-24360
ISBN 0-389-20456-0

CONTENTS

List of Figures
List of Tables
Preface
Acknowledgements

LIST OF FIGURES

LIST OF TABLES

PREFACE

This book is about land. More particularly, it deals with some of the most basic problems which are associated with land. Of these, the most fundamental appears to be the fact that the supply of land in the world is fixed, while the demands upon it seem to be increasing as economic and population growth occur. Others spring from the fact that each parcel of land, or site, is usually contiguous with others, and thus the activities on one may affect the amenity, usefulness and value of its neighbours. As a result of these characteristics the ownership and use of land often carry with them great advantages and power. Indeed, so great have these been, and so easily abused, that it has been felt to be necessary for the state to regulate the aggregate, intensity and spatial pattern of land use in the interests of society at large. Moreover, this need has been felt in many countries, including both some with high and some with low densities of population, some with developing and some with developed economies, and some with free-market as well as those with centrally-planned systems of economic management. It was also felt especially strongly during the late 1960s and early 1970s, and many controls were introduced by governments during that period. However, wide differences in the perception of the land problem, and in the choice of methods for tackling it, have grown up even within the relatively small number of countries which possess developed economies; and it is the purpose of this book to examine both the general form of the land problem and the variety of perception and of ways of dealing with it amongst these countries.

In doing this the book comes close in aim and content to, but is not primarily concerned with, a large number of academic disciplines which are concerned with land. Thus, it is not about land resources, their recognition and evaluation, nor does it deal solely or even largely with land economics - the study of the way in which land is allocated among competing uses - nor it is a text on economic geography, telling of the way in which those uses sort themselves out over space, nor a treatise on land reform or the taxation of real estates, nor a history of land-use planning, though all these topics impinge very closely upon the land problem, and will be considered at least in part in the text. There are many books which deal much more directly with all these topics, some of which are referred to in the following pages and are listed in the bibliography. Rather, this book attempts to describe the way in which the two basic and recurring reasons for concern over land which were mentioned at the outset are perceived and dealt with in four, major developed economies. As such it comes closest in methodology to a long-established, but recently-scorned, form of geographical enquiry - the regional study - in that it seeks to relate the resources of the natural and spatial environment, and the culture of the inhabitants of each country, to the perception of the land problem which is held by each society and the form of land use, and in this case of land-use regulation, which each has adopted.

Such a task requires an intimate knowledge of many parts of the world, and it would be quite wrong to claim that this is what I possess. What information has been collected is largely due to the generosity of the British Council, the Carnegie Trust for the Universities of Scotland, the Jagiellonian University of Krakow, the Royal Society and the University of St. Andrews, which have defrayed the expenses of visits to Eastern Europe, Japan and the United States. It is also due to the selfless assistance and hospitality provided by a very large number of people in many parts of the globe - elected representatives, senior officials of central and local government, academics and businessmen - too numerous to name individually, who have made the work of collection not so much a labour as a pleasure. Nevertheless, this has been a work of exploration. It is not a final statement.

Lastly, I wish to thank my long-suffering cartographer, Mr. Graeme Sandeman of the Department of Geography in the University of St. Andrews, who drew all the figures except 6.1, and my typists - Mrs. Muriel Hendry, Mrs. Betty Niven and Nancy Wood - for coping with my execrable scrawl. Their work has been impeccable - the errors are all mine.

Andrew H. Dawson

ACKNOWLEDGEMENTS

Grateful acknowledgement is made to Universe Books, New York, for permission to publish Figure 2.1 from <u>The Limits to Growth: A Report for the Club of Rome's Project on the Predicament of Mankind</u>, by Donella H. Meadows, Dennis L. Meadows, Jorgen Randers, William W. Behrens III, a Potomac Associates book published by Universe Books, N.Y., 1972, Graphics by Potomac Associates; to the International Federation of Housing and Planning Wassenaarseweg 43, 2596 CG, the Hague, Netherlands, for permission to publish Table 5.2; and to the Planning Director, Framingham Township, Massachusetts for permission to publish Figure 6.1.

Chapter 1

LAND AND THE DEVELOPED ECONOMY

Is there a land problem? If so, what form does it take? Is the finite supply of land on the globe inadequate to support the present population? Or is there a land shortage only in a few countries or in some limited regions? Is land used in a way which represents its true value to the community? Or is shortage the consequence of inefficient operation of the land market, or of the power of landowners? More particularly, is it possible for a developed economy to suffer from a land problem?

Questions such as these are not new. Indeed, they have caused debate ever since Malthus published his _Essay on the Principle of Population_ almost two hundred years ago. Nor will they be answered finally in this book, for it is too much to hope that after such a lengthy period of controversy any single view will command general agreement. The aim here is much more modest. It is to outline the major arguments for and against the proposition that land is in short supply, and that its use must therefore be carefully controlled; to examine the chief participants in land-use decisions which are made within systems of controls; and to describe land-use regulation in a selection of developed countries at the present time. In so doing it is hoped that the continuing debate about land will be better informed. It will not be concluded.

The land problem arises from two fundamental characteristics. In the first place it has long been acknowledged that land is in fixed supply, and that in this it differs from the other factors of production - labour and capital. For this reason some writers, and especially Malthus, have suggested that there is a limit to the population which the earth can sustain; and malnutrition and starvation in primitive societies have been seen as proof of

such a proposition. But advanced economies are not
constrained in this way. They can raise the
productivity of land by means of chemical
fertilisers and high-yield strains of crops. They
can intensify its use through factory farming and
cultivation in greenhouses. They can combat plant
disease with insecticides, applied if necessary from
aircraft; and thus they can raise farm output at a
faster rate than that of population. Alternatively
they can specialise in the production of
manufactured goods and services, and use them to pay
for imported foods. Thus they have the means of
sustaining simultaneously very high densities of
population and standards of living. Indeed, some
developed countries face what appear to be endemic
problems of agricultural overproduction; and many
contain large tracts of both rural and urban land
which have fallen out of use. On the other hand,
the wide range of activities and types of production
in affluent societies makes for a much more varied
and expansive demand upon land than is the case in
developing countries. High standards of living
beget low densities of housing and sprawling urban
settlements, which depend in turn upon the provision
of extensive road systems. They give rise to
demands for such land-consuming recreational uses as
golf courses, and they are based upon extractive and
manufacturing industries which are also major land
users. Furthermore, many of these developments
occur in areas which are used for agriculture, and
especially in lowlands on which the soils tend to be
of better quality. All this has led to widespread
demands, even in the rich nations of the world, for
the protection of prime agricultural land, and also
of areas which are perceived to be of historical,
scenic or scientific importance, from such
developments.

        Secondly, it is also generally appreciated that
land is specific in location, that sites are
immobile, and that as a result their character and
value often owes as much to the use which is made of
neighbouring sites as to their inherent qualities.
Nuisances which spill over on to them may reduce
that value, while the improvement of adjacent plots
may lead to unearned, windfall increases in their
market price. In the developed economy the scope
for such spillovers is very great. Huge, coal-fired
power stations affect not only the sites immediately
next to them, but spread acid rain over thousands of
square miles, to the detriment of both those who
consume their electricity, and those who do not.
Similarly, the filling and draining of wetlands or

marshes may not only destroy rare wildlife habitats, but also increase the likelihood of flooding downstream. However, it is unusual for the costs of these losses to be met directly by those responsible for them. In other words, the market fails to allocate the costs and benefits of such operations accurately; and, because this is the case, there has been an increasing tendency for governments to restrict the rights of private owners to use their land as they wish, in the interests of society at large.

However, the argument has not been all one way. Voices have been raised against such interventions by governments. Some have echoed the views of Locke and Jefferson that land ownership makes the individual independent of the state, strengthens democracy and thwarts the abuse of government power; and they have noted how state ownership and control can lead to corruption and discrimination (Bjork 1980). Others have pointed to the failure of zoning and other planning devices to maintain farming in the urban fringe or to redevelop empty inner-city sites, to the expense involved in the drawing up and implementing of land-use plans and zoning restrictions, to the delays which are imposed on developers by the need to obtain approval before work can start, and to the drab uniformity of regulated development (Sharpe 1975, Strong 1981). Furthermore, the history of economic growth argues strongly for the private and unfettered ownership of land. Replacement of the communal organisation of land use under feudalism in Europe and Japan by private ownership conferred on owners the right to use their land as seemed most profitable, or to dispose of it to those who could use it more successfully, and opened the way to improvements in productivity which not only provided capital for investment in new manufacturing industries, but also food for the growing urban population working in those industries. Some writers have even claimed that there is no foreseeable land shortage in the world, and many have suggested that Man is capable of using the earth's surface much more effectively than at present. In short, there is no agreement as to how best to manage the use of land, nor as to whether Malthus and his supporters will be proved right. Nevertheless, the issue is fundamental to any justification of land-use regulation, and therefore we shall examine the main arguments concerning both land shortage and market failure in Chapter 2.

Whatever the balance of the argument about

whether or not land is in short supply or whether
market failures are sufficient to merit intervention
by the state, almost all developed countries have
introduced some controls over land-use change during
this century. As a result they have opened up the
land-use decision-making process to a much wider
range of individuals and organisations than would
have been involved in a system based strictly on the
market.

In this process central government usually
articulates the needs of society as a whole, while
local government is called upon to carry out much of
the   day-to-day   administration   of   the
central-government's   policy.   However,   the
perspectives and views of local and central
authorities are rarely identical, and conflicts of
interest arise between them. Furthermore, other
interested parties are also able to press their
points of view when public discussion replaces
market decisions; powerful groups may be able to
exert pressure on permit-granting authorities behind
the scenes; and other, less influential bodies may
react by disrupting public inquiries and by
physically resisting the implementation of
controversial proposals. In the absence of
strictly-market decisions, land-use change will
often be a compromise between the site owner, the
developer, local and wider interest groups, and
local and central government; and the result in any
particular case will depend in large measure upon
the relative strengths, and the degree of interest,
of the various participants. In the most extreme
case there is always the danger that regulatory
agencies may become just as much the creatures of
those who they are intended to constrain as market
decisions are determined by those with most money.
The success, or otherwise, of any system of control
will depend in large part upon the extent to which
it enables those without land to remedy the failings
of the market. However, such groups often remain
much weaker than those who they are opposing, in
spite of the intervention of the state. We shall
examine the general types of participant, or
'actor', who become involved when government seeks
to influence land-use decision-making, and the
relations between them, in the developed economy in
more detail in Chapter 3.

However, all this presupposes that the
definition of the 'developed economy' is clear, and
that the identity of countries with such economies
is known. It also suggests that, in all respects
concerning land use and its regulation, all

developed economies are similar. In fact, none of these conclusions would be correct; and therefore we must spend a little time discussing what is meant in this study by 'development', and showing why it will be necessary to examine the pattern of land-use regulation in not one, but several, different developed economies.

Many criteria have been used to measure the degree of development, and few have yielded exactly the same classification of the world's states. One of the most frequently employed has been Gross National Product per capita, but others have been the degree of urbanisation and the levels of literacy or infant mortality. Some of these have an important bearing on the land problem, but others are more tenuously connected to it. Some have been measured on a comparable basis for almost all countries, but information about others is difficult or impossible to acquire. In view of these difficulties an independent measure of development has been drawn up for this study based upon criteria which are available for all the countries listed in the World Development Report, and which are closely related to land issues.

Of these, Gross National Product per capita is one of the most important. Affluent nations make heavy demands upon land for housing and recreation - often having a large number of second homes - and for transport facilities. They can usually afford a high degree of mobility, either in the form of public transport or, more likely, the widespread private ownership of cars; and their populations have time to worry about such land-use issues as the preservation of historic landscapes or the habitats of rare fauna and flora, which are not directly or immediately related to their survival. Secondly, developed economies are, by definition, broadly based. They contain large secondary and tertiary sectors, and do not exhibit any narrow dependence upon primary forms of production. The country-to-town migration is largely complete, and a large proportion of the population lives in, or is in close contact with, the system of urban settlements. Indeed, 'getting out of town' is often more important to inhabitants of the developed economy than seeking 'streets paved with gold'. Thirdly, developed countries have high educational and medical standards. Rates of fertility are low, population growth is slow, levels of life expectancy are high, and population pressure on the land, at least insofar as it is directly linked with the threat of starvation, is absent.

Table 1.1: A Listing of Countries according to their Degree of Development

(1) The Most Developed Countries

	GNP per capita 1978	% employed outwith agriculture 1978	% of population in towns 1980	Life Expectancy at birth 1978	Fertility 1978	Sum of rankings	Overall Rank or level of development
Sweden	3	8	7	2.5	6	26.5	1
Denmark	4	14.5	10.5	5.5	6	40.5	2
United Kingdom	19	2.5	2	12	6	41.5	3
West Germany	6	7	8.5	21.5	1	44	4
Netherlands	10	10.5	20.5	5.5	3	49.5	5
U.S.A.	5	2.5	15	12	11	53.5	6
Canada	8	10.5	15	5.5	15	54	7
Australia	12	10.5	4.5	12	16.5	55.5	8
Switzerland	2	10.5	40.5	5.5	2	60.5	9
Japan	14	22.5	16.5	1	11	65	10
France	11	16	16.5	12	15	70.5	11
Belgium	9	5.5	24.5	21.5	11	71.5	12
New Zealand	20	17	8.5	12	19	76.5	13
Norway	7	14.5	48.5	2.5	11	83.5	14.5
Singapore	29	2.5	1	34.5	16.5	83.5	14.5
East Germany	18	17	18.5	21.5	11	86	16
Hong Kong	34	5.5	3	21.5	31.5	95.5	17
Italy	22	22.5	27	12	15	98.5	18
Israel	25	12	4.5	21.5	38.5	101.5	19
Austria	15	17	46	21.5	6	105.5	20.5
Finland	17	24.5	36.5	21.5	6	105.5	20.5

(2) The Next Most Developed Countries

						Sum of rankings	Overall Rank
Spain	26.5	28.5	22	12	31.5	120.5	22
Czechoslovakia	21	20	35	34.5	25.5	136	23
Uruguay	42	20	10.5	28	36.5	137	24
Argentina	39	24.5	13	28	36.5	141	25
U.S.S.R.	23	27	32	34.5	25.5	142	26
Ireland	26.5	31	40.5	12	38.5	148.5	27
Greece	30	48	36.5	12	22.5	149	28
Hungary	28	28.5	46	34.5	19	156	29
Poland	24	42.5	42	28	22.5	159	30
Bulgaria	31	50.5	34	21.5	22.5	159.5	31
Taiwan	47	46	18.5	21.5	28	161	32
Kuwait	1	2.5	6	42	111	162.5	33
Chile	46	31	14	45.5	34	170.5	34
Venezuela	35.5	31	12	47	53.5	179	35
Cuba	67	34	32	21.5	28	182.5	36
Jugoslavia	36	42.5	64	42	19	203.5	37
Panama	48.5	44.5	46	34.5	44	217.5	38
Costa Rica	44	38	63	34.5	40	219.5	39
Trinidad & Tobago	35.5	26	96	34.5	31.5	223.5	40
Jamaica	53	37	55	34.5	45.5	225	41
Portugal	38	35.5	83	42	28	226.5	42

Notes: (1) Numbers represent the rank order of countries. That with the highest value is ranked one, except for 'Fertility' and the 'Overall Rank', based on the 'Sum of rankings', where the lowest value is ranked one.

(2) For all except fertility the first rank was awarded to the highest value. In the case of fertility it was to the lowest value.

(3) Countries with overall rankings of 43 or over are omitted from the table.

Source: World Bank, World Development Report 1980, Washington 1980, Tables 1, 18, 19, 20, 21.

Each of the 125 countries listed in the Report
has been ranked according to those of the three
types of characteristics listed above for which
information is available for all countries, and a
composite index of development has been calculated
by summing the scores for each country (Table 1.1).
The countries have been divided into groups of
twenty-one, so that each group represents one sixth
of the countries considered. Those ranked one to
twenty-one may be considered to have the
'most-developed economies', and those ranked
twenty-two to forty-two the 'next-most-developed' in
the world. Inspection of the list reveals the
absence of a number of small countries which are not
included in the Report, but which might have merited
inclusion in the 'most-developed' sextile, such as
Iceland and Luxembourg. It also shows that the
classification is similar, but not identical, to
those produced by some other studies of
development. The countries in the 'most-developed'
category are much the same as those recognised by
the Pearson Commission (1969) and the World Bank as
'industrialised' or 'developed market economies',
with the exception of South Africa, which is ranked
forty-seven on the criteria and method employed to
contruct Table 1.1. Countries in the
'next-most-developed' sextile are among the most
prosperous of the 'developing countries' in those
lists, and include all of the 'centrally-planned
economies' in Europe with the exception of East
Germany, which is in the first sextile, Albania and
Romania. However, it should be noted that several
of the Latin American countries included in this
group are not normally considered to belong to the
developed world. On the other hand, it is relevant
to note that, according to one important measure of
development - the supply of food as measured by
calories per capita - these countries have all
enjoyed a level of nutrition in recent years which
has been equal to, or in excess of, that which is
necessary, whereas many of those which do not fall
into either the 'most-developed' or the
'next-most-developed' groups have suffered from
inadequate food supplies (FAO 1982).
    However, some important differences do exist
between the countries which have been defined as
developed by this system of classification, one of
which is the amount of land available per capita of
the population; and for this reason all of the 125
countries have also been ranked according to
population density, and divided into three groups of

Table 1.2: The Developed Countries by Population Density

(1) Most highly developed countries

   (a) Very densely populated

        Israel, Hong Kong, Japan, Singapore, Belgium,
        Italy, Netherlands, U.K., West Germany

   (b) Densely populated

        Austria, Denmark, East Germany, France,
        Switzerland

   (c) Of medium or low population density

        Australia, New Zealand, Finland, Norway,
        Sweden, Canada, U.S.A.

(2) Highly developed countries

   (a) Very densely populated

        Taiwan,
        Jamaica, Trinidad and Tobago

   (b) Densely populated

        Czechoslovakia, Hungary, Jugoslavia,
        Poland, Portugal,
        Cuba

   (c) Of medium or low population density

        Kuwait,
        Bulgaria, Ireland, Spain, U.S.S.R.,
        Costa Rica, Panama,
        Argentina, Chile, Uruguay, Venezuela

Ranges of values:

   Very densely populated: above 165 inhabitants per square
   kilometre

   Densely populated: between 80 and 165 inhabitants per
   square kilometre

   Of medium or low population density: below 80
   inhabitants per square kilometre

'very-densely-populated',   'densely-populated'
countries and countries which are 'of medium and low
population density'. The countries in Table 1.1 are
shown in their appropriate groups in Table 1.2.
Examination of this table reveals that there are
several instances of countries of similar
geographical location, history or size occurring
together in the various sub-groups.

The most interesting sub-group from the point
of view of this study would appear to be (1a), that
is, the sub-group which combines both very high
levels of economic development and population
density. Countries in this sub-group are either in
north-west Europe or Asia, and many are of very
small area. Some, such as Hong Kong and Singapore,
are little more than city states, and are therefore
of limited interest to this study, for there is no
possibility of them having a significant
agricultural or forestry element in their
economies. But others, such as Italy, Japan, the
United Kingdom and West Germany, are of much greater
interest because they are of medium size in both
area and population, possess large rural tracts, and
have rural industries which compete for land with
urban development. Both the United Kingdom and
Japan will be examined in more detail in Chapters 4
and 5 respectively. At the other extreme lies
sub-group (1c), containing those countries with the
most highly-developed economies, but which have only
medium or low population densities. These are to be
found in high latitudes (Canada, Finland, Norway and
Sweden) or among those countries which have been
settled largely by English-speaking peoples in
modern times. (If South Africa had been included in
the 'most-developed' group it would also have fallen
into this sub-group.) The largest and most important
of these is the United States of America, and it
will be the subject of detailed study in Chapter 6.
Amongst the 'next-most-developed' economies those
which fall into sub-group (2a) are not only few, but
also all rather small in both area and population.
Sub-group (2b), in contrast, is made up almost
entirely of centrally-planned economies in Eastern
Europe. Furthermore, it contains most of the
countries with this type of economy. Those which
are not included in the sub-group are either
somewhat more developed, such as East Germany, or
have a lower population density, such as Bulgaria or
the U.S.S.R. In other words (2b) is the modal class
for European centrally-planned economies, and for
this reason its members, and especially the

largest - Poland - will be examined in Chapter 7.
Thus the countries which have been chosen for
detailed examination, although all economically
developed, have been drawn from groups which vary
widely in population density or economic system.

It should also be noted that they differ from
one another in other characteristics which are
relevant to the question of land. For instance, the
high density of population of Japan is exacerbated
by the mountainous character of the archipelago, in
which three-quarters of the surface is in slopes of
more than fifteen degrees, while over sixty percent
of Poland and the United Kingdom are low lying,
gently sloping and have climates and soils which are
suitable for agriculture. The United States, in
contrast, covers a vast area including not only the
Rockies and the deserts of the western states, but
also enormous tracts of land of high agricultural
potential in the Midwest and the sub-tropical
South. These contrasts are reflected in the very
wide differences in the proportions of these
countries which are devoted to agriculture and
forestry, and in the amount of agricultural land per
capita (Table 1.3). They are also an important
reason for the fact that, while Japan is one of the
largest food importers in the world today, the
United States is the chief food exporter.

Secondly, recent studies have revealed that
there are also great contrasts in the way in which
land issues are viewed in these countries. Whereas
in the mid-1970s Americans saw the growth of
population, the supply of energy and minerals, and
the problem of run-down urban neighbourhoods as
serious environmental threats to the future, many
fewer Japanese did so in spite of the much higher
population density and lower level of mineral
resources in their country, though they did place
the preservation of the natural environment high on
their list of desirable aims. Conversely, many more
Japanese felt that more should be done to control
and plan the impact of demographic and economic
growth than Americans, even if this would restrict
the rate of economic development. Thus, in spite of
their high standards of living, Americans gave
greater priority at that time to further economic
growth than to environmental protection (McGlen,
Millrath and Yoshi 1979).

Another study covers the agricultural support
policies of governments. Bale and Lutz (1981) have
shown that the farming industries in many of the
developed economies, including Japan and the United
Kingdom, have enjoyed a substantial degree of

Table 1.3: Population and Land Use in the Four Countries

	U.K.	JAPAN	U.S.A.	POLAND
Area (square kilometres)	24,482	37,230	936,312	31,268
Population in 1979 (thousands)	56,142	115,870	220,286	35,257
Agricultural land in 1979 (% of total)	76.3	14.8	47.2	62.4
of which Arable and permanent crops in 1979 (% of total)	28.6	13.2	20.6	49
Woodland in 1979 (% of total)	8.6	67.4	31.9	28.5
Agricultural land per capita in 1979 (hectares)	0.33	0.05	1.95	0.54
Population growth in 1961-79 (%)	6.1	23.2	19.9	17.7
Change in percentage of land in agriculture and woodland 1961-79	-4	-5.2	-2.4	-1.8
Change in percentage of land in arable and permanent crops 1961-79	-1.7	-3.1	0.3	-2.7

Source: FAO Production Yearbook 1981, Rome 1982, tables 1, 3.

protection in the recent past, unlike the situation in developing countries where governments tend to tax farmers heavily and thus depress output. Nevertheless, it should also be noted that there has been a very wide range amongst the levels of that support between the developed countries. While at least two-fifths of the cereal crops of Japan, including rice, may have been produced solely in response to the very high, guaranteed prices which were on offer from the government, only a small proportion of the United Kingdom's cereal harvest could be explained by such intervention in the market.

On the other hand, the four countries in Table 1.3 also display important similarities which touch upon the question of their land resources. All have experienced an increase in population during recent decades, although that increase was much smaller in the United Kingdom than in the others. Similarly, all have undergone a decline in the area of land given over to agriculture and forestry, although that of cropland in the United States did rise.

Put briefly, the activities of government in the matter of land use in these four countries have taken place against complex and contrasting political, demographic, physical and cultural backgrounds. We shall attempt to compare the success of otherwise of those activities in the final chapter when we have examined those backgrounds in more detail.

But before we begin to make this comparison one other matter should be considered, namely, why is it appropriate to undertake such a study at this time? In order to answer this question is is necessary to note that interest in the land problem has not been constant or steady during the last hundred years. Crises have arisen, such as the submarine blockade of Britain during the Second World War or the food and oil shortages of the early 1970s, which have suddenly focussed public attention on land-use issues, and led to important extensions of government regulation of land-use change. At other times, in contrast, there has been little interest in these matters. Some writers (Boyer 1981, Edel 1981) have explained this history in terms of the long cycles of investment in the economies of developed countries, which have been identified by Kondratieff (1935) and Rostow (1975). They have argued that, whereas there is little or no interest in the imposition of new, restrictive, land-use controls by governments during periods of weak investment, economic depression and unemployment, nor in the years of accelerating investment and economic growth which follow, the situation is very different in the later stages of these cycles. They suggest that the decline in the yield from investment in productive activities at the peak, and during the later stages, of the cycle brings forth demands for government intervention to protect the profitability of these activities, and to constrain the profits which are being made from the rapid rise in land prices, which follow from the diversion of investment from productive activities into the

ownership of real estate. Those who support this
explanation of the periodicity of interest in
land-use controls can point to the increase in the
zoning of land in American cities in the 1920s, and
the imposition of new controls in Japan and many
other countries in the late 1960s and 1970s, as
evidence for their view. However, it is also
possible to interpret this pattern as a response to
public concern about the rate of development, and of
the associated land-use and landscape change, at the
point in the cycle when they are most rapid.
Whatever the reason, a considerable increase in
land-use regulation did occur during the 1960s and
1970s in almost all the developed economies,
including those of Eastern Europe, and this has been
replaced in the late 1970s and 1980s by a greater
emphasis on the need for economic growth, and by
demands for the relaxation of government controls.
If the long-cycle model is correct, the next peak of
investment, and wave of concern over land use, may
not occur until the early years of the twenty-first
century. Little new legislation is in prospect for
the 1980s, but few statutes have been repealed; and
many which were enacted in the 1970s are now being
implemented. Thus, this may be a relatively quiet
and stable period in which to summarise and compare
the extent to which different economies within the
developed world have responded to the land problem.

Chapter 2

MALTHUS AND HIS FOLLOWERS

"Geographers, economists and others have been
announcing at regular intervals for a long time
that the world's cultivatable land is
exhausted." Grigg, D. (1981), The
historiography of hunger: changing views on the
world food problem 1945-1980, Transactions
I.B.G. 6, p. 285.

INTRODUCTION

In the first chapter we touched on several
aspects of land use in developed economies. In
particular, we noted that there is a wide variety of
competing uses; and that, despite equally wide
variations in the amount of land per capita amongst
these countries, land-use changes in a selection of
them have shown similar trends. But, as will be
seen in more detail in later chapters, attitudes
towards land and its use have also varied widely in
the recent past; and the countries with the smallest
land resources have not necessarily been either the
earliest or the most stringent in the controls which
they have imposed on them. Nevertheless, there has
been a widespread movement of opinion in developed
countries since the mid-1960s in favour of greater
public regulation of land use; and this has renewed
a debate which has been in progress since the days
of Malthus. It is the purpose of this chapter to
review briefly the case which was put by Malthus,
and that which has been argued in a more complex
form in recent times by his followers, before
considering whether or not these or any other
circumstances justify the regulation of land-use
change by the state.

MALTHUS

     In his __Essay on the Principle of Population__
(1798) Malthus argued that society is constantly
facing a potential land crisis. This arises because
of the ability, and propensity, of the human species
to reproduce itself rapidly, and to grow at a
geometric rate, while the output of agricultural
products can only increase, at best,
arithmetically. In other words, while the output of
farm products is growing from two to four to six to
eight, population is increasing from two to four to
eight to sixteen. Given abundant natural resources,
Malthus noted that the human race may double in as
little as twenty-five years, but he warned that
eventually it will outrun those resources, and will
then be subject to such painful checks as war
between those scrambling for what has become scarce,
malnutrition, which will reduce resistance to
disease, and, ultimately, starvation.
     These conclusions were based largely upon
evidence from the United States and the United
Kingdom. Malthus had noticed that in the
late-eighteenth century the United States was
offering what seemed to be inexhaustible resources,
and that its population had doubled in the space of
about thirty years. He had also noted that, because
of the youthfulness of that population, it was
likely to double again within a similar period in
the early-nineteenth century; and he was of the
opinion that the population of the United Kingdom
likewise was increasing rapidly, although not quite
at the United States' rate. Turning to agricultural
production he acknowledged that there had been
substantial, and indeed revolutionary, improvements
in farming in the United Kingdom during the
eighteenth century, which had led to a substantial
growth in output. That further increases were
probable Malthus did not doubt, but he argued that
it was unlikely that the eighteenth-century
improvements could be more than matched in the
future. As a result, he forecast that, while the
population of the United States could continue to
grow for many years, future increases in the United
Kingdom would precipitate a crisis before the end of
the nineteenth century unless the minimum age for
marriage was raised, and the level of fertility
reduced (pp. 54-110). In fact, population grew
from 10,501,000 in 1801 to 37,000,000 a century
later; and at the latter date people were better
clothed, fed and housed than when Malthus had

written.

Why did the disaster not occur? With the benefit of hindsight it it relatively easy to discern the immediate reasons. Technical innovation in the United Kingdom increased labour productivity, as did increasing specialisation among workers; and free trade enabled the country to grow wealthy as it concentrated upon the production of those goods in which it was competitive, and the import of those for whose output it was less suited. Moreover, these changes were accompanied by the demographic transition, as the falling death rate of the period between the 1860s and 1920s was followed during the 1870s to 1930s by diminished fertility (McKeown 1976, p. 29). Indeed, fertility declined to the extent that, between the two World Wars, it appeared that a natural decline would occur in the British population. Thus, just at a time when, according to Malthus, conditions were becoming ever more favourable for fertility, the birth rate was falling; and this transition was taking place not only in the United Kingdom, but also in the other industrial economies of Western Europe and North America. Thus nineteenth-century Britain - but not Ireland - escaped the Malthusian threat.

However, it is not so much the immediate reasons for that escape as the more general causes which are of lasting significance for an evaluation of Malthus' theory, for the Industrial Revolution in Europe and the development of that continent's world-wide trade, based upon the specialisation of production, were specific events in history, which may not be repeated. The simultaneous increase in both population and the standard of living was explained more generally by Chadwick in 1888 (p. 6) when he claimed that, as population increases, so does the state of knowledge. In this he shared the view of Engels, that it is in the nature of Man to consider the problems facing him, and to react to them in a reasoned manner in order to avoid or minimise their impact (Meek 1953, p. 63). This argument was extended further by Hoyle (1963), who suggested that, because increases in population encourage innovation and improved productivity in the use of natural resources, they carry us further away from the moment of Malthusian crisis, rather than towards it - a claim which appears to be borne out by the history of land use in Britain in the nineteenth and early-twentieth centuries. It is ironic that the inventions of that period, which permitted cheap grain, and later meat and dairy products, to be imported into the United Kingdom

Table 2.1: Land Use in Britain 1871 - 1939

Percentage of improved land in

	Arable	Permanent Pasture
1871	59.7	40.3
1901	48.1	51.9
1939	40.6	59.4

Source: Ministry of Agriculture, Fisheries and Food,
         A Century of Agricultural Statistics, HMSO, London
         1968, pp. 94, 96.

from the Americas and Australasia, and which
benefitted almost everyone in the country, should
have caused distress only to the farming community.
As prices for agricultural products fell, land
'tumbled down to grass' (Table 2.1), and on the
edges of the hills fields were abandoned. By the
1930s there was widespread evidence to suggest that
the United Kingdom had more than enough land.

NEO-MALTHUSIANISM

But has the crisis merely been delayed? Whereas
approximately one billion people were alive in the
world at the start of the nineteenth century, the
number had risen to three billion by 1960, and the
estimate for 1980 was 4.4 billion. This
accelerating growth has occurred because the world
does not only contain developed countries, which
have experienced the demographic transition, and now
have slowly-growing populations, but also huge
numbers of people in the relatively poor, developing
economies. Indeed, the vast majority of mankind is
to be found in this second group of countries; and,
because the population there is both much larger
than in Malthus' day and young, the absolute
increases in the world's population are very large.
For example, that in the single year of 1980 was
about 77,000,000 (United Nations 1983, table 1); and
most forecasts suggest that there will be more than

six billion people alive in the year 2000. Moreover, it is unlikely that the world population will cease to grow before the end of the next century, when it has been estimated that it will be between eleven and fifteen billion, if catastrophe does not intervene (Leszczycki 1980, p. 96).

Growing awareness of these facts since the Second World War has given rise to neo-Malthusian fears concerning the capacity of the earth to feed such a huge and rapidly-increasing number of mouths, even if there is the most careful conservation and protection of all the land which could be used for farming. Various estimates have been made since the 1960s of the carrying capacity of the globe, but they differ so widely that all must be treated with care. At one extreme the Ehrlichs (1970, p. 3) have suggested that four billion is already too great for the maintenance of a permanently-sustainable human occupation of the planet, while Malin (1976, pp. 19-20) has claimed that 290 billion could be fed. Amongst the less extreme forecasts, one of the most influential has been the work of Meadows' team, entitled The Limits to Growth. Interest was aroused not only by the timeliness of its publication in 1972, in a period of severe food shortage in some parts of the world, but also by the ideas and methods which it employed, and especially the concept of negative feedback. By using this the authors showed how population and economic growth could actually reduce the ability of the earth to produce food, by transferring land from agriculture to housing, industry, and transport, by increasing the pollution of the atmosphere through the growth of industrial output, and thus exacerbating such problems as acid rain, the leaching of soils and the killing of forests, and by the rapid exhaustion of the mineral resources upon which advanced, high-yielding agriculture has come to depend for fertilizers, fuel and pesticides. Thus they demonstrated that the impact of rapid economic growth upon the food-producing potential of the earth was not linear, but compound. Many factors were included in the study, and they and the links between them, endowed the conclusions with an impression of comprehensiveness. Moreover, the application of the computer to the calculation of the exact date at which these inter-related factors would cause a major collapse in society gave the results an air of accuracy and certainty which, at first sight, was very frightening.

The Limits to Growth concluded that the area of land which is potentially suitable for agriculture would soon prove to be inadequate. This area was taken to be 3.2 billion hectares, while the area required to feed one person at the standard of nutrition in the United States was assumed to be 0.9 hectares. These assumptions were then used to

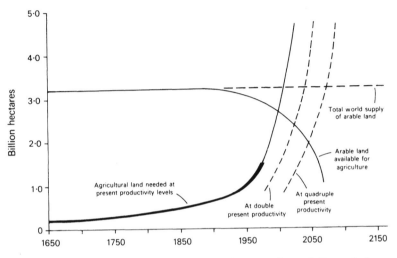

Figure 2.1: The Meadows' Model of the Supply and Demand for Arable Land in the World

produce Figure 2.1, which shows that, using the then-current population forecasts and level of agricultural technology, a world food crisis, in which the supply of land was inadequate to feed the population, would occur in the 1990s. It shows also that, if the productivity of land could be doubled, the crisis would be postponed until about the year 2025, and that, if it could be quadrupled, it would come in about 2045. However, if arable land could also be protected from development, the crisis could be delayed until 2010, 2040 or 2070 respectively, But, whatever efforts were made both to protect farmland and increase its productivity, the world's population was forecast to rise swiftly to sixteen billion, and at that point overshoot the earth's resources and collapse to a much smaller figure as Malthusian checks came into operation. Furthermore, it was suggested that it was too late to avoid such a catastrophe.

However, the situation may not be quite as grim as the Meadows team have suggested. Just as Malthus may have been misled by the accelerating growth of population in late-eighteenth-century England

(Edwards and Wibberley 1971, p. 2) and the United
States, so <u>Limits</u> may be based on what will
subsequently be proved to be an overestimate of
world population-growth rates. The demographic
transition in developing countries appears to be
proceeding faster than was the case in the developed
economies; and it is expected that birth rates will
fall rapidly in the 1980s and 1990s. As a result,
estimates of the world's population by the end of
the century were falling throughout the 1970s; and
even the figure of six billion may only just be
reached (World Bank 1979, p. 183). Nor does it now
appear to be likely that the population will
eventually reach the level at which it will
overshoot the earth's resources. In a more recent
study, Mesarovic and Pestel (1975) have concluded
that catastrophes are much more likely to occur at
the scale of the sub-continent, rather than in the
world economy as a whole. On the other hand, they
may take place before the middle of next century in
places where the supply of cultivatable land is
already almost exhausted, and the population is
engaged in a subsistence or near-subsistence
economy. The most vulnerable area, in their view,
is Southern Asia, where the amount of cultivated
land per capita is much below the average for the
developing economies, where a much larger proportion
of the land which is suitable is already cultivated,
and where the area per capita will decline at an
accelerating rate until at least 2025. Catastrophe
in that region might "start in the early 1980s and
peak around 2000: deaths related to food shortage
would be double the normal death rate"(p. 123).
But they do not believe that this is inevitable if
rapid economic development is achieved with the help
of the wealthy nations; and Barna (1979, pp.
96-111) has confirmed more recently that the
developed market economies should be able to supply
any temporary shortage of food in the rest of the
world up to at least the end of the century.
    What is more, the land resources of the world
may not be as limited as the Meadows' study
assumed. The International Biological Programme,
which was carried out for UNESCO between 1964 and
1975, has produced a new estimate of 3.7 billion
hectares of cultivatable land, half of which could
be irrigated. If only two-thirds of this area were
to be used, as against the thirty-seven percent of
it which was in use in the mid-1970s (FAO 1981, p.
45), and only two-thirds of the maximum yields of
grain were obtained from it using known techniques,

the world grain output could be thirty times the
present level (Buringh et al. 1975). Nor is the
possibility of a substantial increase in doubt for
Bale and Lutz (1981) have drawn attention to the
fact that many developing countries have adopted
taxation policies which depress agricultural output;
Chou and others (1977, p. 9) have claimed that the
real problem is not one of a shortage of land, but
of the institutional arrangements for its
management, and the efficient and equitable
distribution of its products; and Beckerman (1974,
pp. 215-240) has argued that an increasing demand
for food will encourage farmers to intensify the use
of land, that increases in the real prices of
agricultural products will attract investment by the
industry, and that new forms of production will
emerge. These new forms were anticipated by
Boddington (1973), who pointed to the potential
which exists for the substitution of microbial and
synthetic products from laboratory and factory for
field and forest crops; and, more generally, Simon
(1981, p. 225) has claimed that economic
development can cause an absolute decline in both
the quantity of labour used per unit area of land
and the area required for agriculture, even in
countries with growing populations, as it has done
in the United States, and that therefore the growth
of population represents not a drag upon such
economic development, but the generation of the
resources by which it can be achieved. Moreover,
Kahn (1976) has suggested that, as the whole world
is made of minerals, we shall be unlikely to exhaust
the resources available to us for the maintenance of
the current, advanced methods of farming or for the
development of even more productive ones. Thus
these writers believe that the feedback effects of
population and economic growth are both more complex
and less rigid than was envisaged in The Limits to
Growth, and that the human race has both the
physical resources and the ingenuity to avoid the
Malthusian catastrophe.
    However, neo-Malthusians, writing since some of
these claims were made, have not been persuaded.
They have suggested that much of the land which
could be cultivated must be inferior to that which
is already in use, and that therefore any expansion
of the cultivated area will be accompanied by
diminishing returns. They have claimed that much of
the land would be far too costly to bring into use,
especially if it has to be irrigated. They have
pointed to the fact that crops are appearing, such
as those from which motor fuel can be made, which

Table 2.2: World Agricultural Production per capita 1961-80

	1961	1965	1970	1975	1980
World	98	101	104	106	107
of which					
Developed market economies	96	102	105	114	119
of which					
Japan	96	101	108	110	95
UK	94	104	110	114	131
USA	98	101	101	113	115
USSR and European centrally-planned economies	99	103	117	118	119
of which					
Poland	104	101	108	117	99
Asian centrally-planned economies	96	105	109	113	118
Developing market economies	99	99	103	102	103

1961-5 = 100

Source: FAO Production Yearbook 1972-1981, Rome 1973-1982, tables 6, 7

will compete for land (Brown 1980), and that the further development of farming in all countries may be restricted by increases in the real prices of fertilisers and fuel, as petroleum and some other minerals approach exhaustion. Nor have they accepted that the stock of agricultural land will necessarily expand, for, as Eckholm and others have indicated, much is being 'lost' as a result of desertification, erosion and the accumulation of toxic salts in the soils of some irrigated areas

Table 2.3: World Food Production per capita 1961-1980

	1961	1965	1970	1975	1980
World	98	101	106	108	109
of which					
Developed market economies	96	102	108	118	122
USSR and European centrally-planned economies	99	102	117	117	118
Asian centrally-planned economies	96	104	108	112	116
Developing market economies	100	99	102	104	105

1961-5 = 100

Source: FAO Production Yearbook 1977-1981, Rome 1978-1982,
         tables 6, 7

(Brown 1978, Eckholm 1976).
      It is, of course, impossible to adjudicate with
certainty between these two sets of highly
speculative arguments. However, it is possible to
examine the development of agriculture world-wide in
recent years to establish the extent to which it
accords with either of them. Since adequate data
became available in the 1950s the output per capita
of both food and all agricultural products in the
world has increased (Tables 2.2 and 2.3). In
developed economies, both of the centrally-planned
and market types, relatively rapid increases have
been achieved in spite of the falling area of land
in agriculture. In North America and Western Europe
improvements in land productivity have more than
offset the abandonment of some of the poorer,
marginal farmland, as well as the transfer of prime
land to non-agricultural uses. Indeed, they have
threatened, through the production of surpluses, to
depress farm incomes. Thirteen of the twenty-one
'most-developed countries' in Table 1.1 had a
smaller area of arable land in 1979 than ten years

earlier, but all except two - Hong Kong and Japan -
had raised their agricultural output per capita over
that period. Similarly, the arable area in eleven
of the 'next-most-developed countries' declined
during the 1970s, but only two of these - Poland and
Portugal - have suffered a fall in the output of
agricultural products per capita. Increases have
also been achieved in the developing countries
despite rapid population growth, and those in the
Asian centrally-planned economies, including China,
are particularly noteworthy. But the progress of
the developing countries with market economies has
been slight, and in many of the African states there
has been a fall in per capita production (FAO
1972-1981). Moreover, at least 500,000,000 people
in Africa, Latin America and Southern Asia were
suffering from undernutrition between 1972 and 1974
(FAO 1977, pp. 53, 127-8); and there was widespread
famine in the Sahel of Africa in the late 1970s. In
short, although average standards of nutrition in
the world have improved rapidly since 1950, and the
severity of famines has been reduced (Grigg 1981,
p. 279), it is by no means certain as to whether
these trends will continue in the immediate future,
or be reversed.

THE NEO-MALTHUSIAN THREAT AND LAND-USE REGULATION IN DEVELOPED ECONOMIES

It may appear that these conclusions are of
most immediate and direct relevance, not to the
developed economies, but to the poorest countries,
of the world. Developed nations, by contrast, are
wealthy, and have generally been able to command
sufficient of the world's supply of food in
international markets without great financial
hardship to their populations. They have low rates
of population growth, and a successful record of
increasing agricultural productivity and output per
capita. Nevertheless, the neo-Malthusian threats
associated with economic growth, and the dependence
of many poor nations upon some of the developed
economies for food, may have serious implications
for the land policies of those economies. For
instance, it may be strategically unwise for
developed nations to become heavily dependent upon
imports of essential materials at any time, as the
United Kingdom discovered at the start of the Second
World War, and least of all when the general demand
for them is rising and the supply is controlled by a
few countries, or when the surpluses which are

available for export are falling, as Western Europe
and Japan found out in 1973 in relation to Middle
East oil. Conversely if, like the United States,
the country is a major exporter of farm products, it
will almost certainly wish to retain not only the
income from those sales, but also any influence
which they may bring over the countries of purchase
(Batie and Healy 1980, pp. 6-11). Moreover, in the
competition between agricultural and other land
uses, only extraordinarily high prices for land,
such as those around some of the cities in Japan,
will stop its conversion to housing and industry;
and, in the absence of government restrictions on
such conversions, only exorbitant prices for farm
products will give rise to such land prices.
However, there may be a limit to the rise in food
prices which the public is willing to tolerate,
whether it be the result of increasing land values
or of some official policy to encourage farmers to
use the declining stock of agricultural land more
intensively, as the Polish government discovered in
1970, 1976 and 1980. Thus, there are many reasons
why the governments of developed economies should
intervene to prevent the transfer of land from
farming, despite the fact that such action may
diminish the scope for the international
specialisation of production, and for international
trade, and consequently lead to a lower overall
level of output and wealth world wide.
    What is more, similar arguments can be used to
justify the regulation of the use of other types of
land. For example, the Forestry Commission in the
United Kingdom was established in 1919 in order to
provide a strategic reserve of timber; and the
continued expansion of its planting in recent years
has been justified on the grounds of import
replacement and the provision of a secure supply at
a time when the availability of timber for export
from other countries seems likely to decline. Some
governments also consider that it is desirable to
prevent building, or other development of land,
where this would 'sterilise' mineral deposits which
might be needed at some future date. Furthermore,
as societies become richer, they generally develop
much wider interests, which give rise to demands for
the protection of sites which might yield
information of archaeological, ecological, historic
or scientific interest, and for the conservation of
areas of great scenic beauty which provide
recreation or wilderness experiences; and
governments have felt obliged to protect land which
offers these resources, as well as that on which

food can be produced. Indeed, in some developed
countries the protection of the potential for food
production has been but a minor land issue amongst
others. What is of interest, however, is that many
of the controversies over land use, and many of the
detailed justifications for regulation by government
which have been outlined above, can be traced to a
single, basic complaint, namely, that a free market
in land amongst private owners is not able to give
full, or even any, expression to the value of land
to society as a whole, and that therefore it fails
to allocate land among the competing uses in a
proper manner.

MARKET FAILURE

One major cause of this underlying market
failure is the difference between the discount rates
of private landowners and society. Individuals who
are faced with a need to make a living, or firms
which are under pressure to make profits, may well
consider that conversion of farmland in their
ownership to housing, industry or roads, the
destruction of rare marshland habitats or
archaeological sites in the pursuit of larger
outputs from their farms, or the mutilation of
beautiful scenery through the exploitation of
minerals are unfortunate, but inevitable. Moreover,
any individual or firm may argue that their actions
are so small as to make little difference to the
total stock of farm, scenic or other land. However,
society as a whole may view the matter very
differently. If the growth of population will
eventually require the use of every suitable piece
of land for food production, or if science will need
the knowledge which can be obtained only from the
study of unique wildlife habitats, no further
conversion of, or damage to, such land can be
permitted by society in the interests of future
generations. In other words, the time horizon of
society is almost infinitely long, and communities
cannot allow that which is irreplaceable to be
consumed by the present generation if the effect is
to reduce the options which will be available to
later ones. Alternatively, we can say that the
resource costs involved in the use of these assets
are much greater than the money costs which are paid
by the users in the market. However, the choice is
rarely between very rapid land-use change and an
absolute ban on it, for, as Irland (1979, p. 52)
points out, the legacy of any generation to those

following includes not only the natural resources
which it has not consumed, but also the stock of
capital - factories, roads and schools - and of
scientific and technical knowledge, which allows
these resources to be used sparingly and
efficiently. Because this is the case, it is
justifiable to hand on, for example, not the entire
stock of agricultural land which has been received,
but a diminished stock together with improved
methods for its use, and only in the case of unique
land resources might their transfer to other uses be
forbidden. Unfortunately, because of its
generality, this rule is of little help; and the
exact area which should be preserved at any time
continues to be a matter of judgement rather than of
exact calculation according to generally-agreed
principles (Irland 1979, pp. 51-52).

A second major cause of market failure in
relation to land arises from spillovers or
externalities associated with its use. Much land is
used not only by its owners, but also by others who
have no formal property rights to it; and, where
these unformalised uses occur without the permission
of the landowner or recompense to him, or in
circumstances in which he cannot collect an
appropriate payment, the market has failed. Thus,
when a factory-owner deposits soot on the washing
hanging in nearby gardens, he is getting rid of his
waste products for nothing, and the owners of the
washing are obliged to go to all the cost and bother
of cleaning their clothes again. Similarly, where
the expansion of a town raises the value of
undeveloped land around it, and local authorities
are obliged to pay high land costs in order to
provide public housing, playing fields and schools,
the landowners are receiving an unearned capital
sum, and the market has failed to internalise the
costs and benefits arising out of land-use change.
If such types of external costs are the consequence
of the actions of a large number of individuals and
firms, each of whose contributions to the total is
small, it may not be worth the while of those who
suffer the consequences to seek redress through the
courts; and it may be more appropriate for
government, acting on behalf of the whole community,
to insist that those who cause the negative
spillovers either pay for them, or behave in some
way which will minimise their effects, or simply
prevent them from occurring. For instance,
factories and houses may be separated by the zoning
of land in order to reduce the effect of atmospheric
pollution, industrial noise and heavy traffic on

residential areas; and farmland may be purchased at
current-use value by government, and subsequently
sold at market value to developers with a view to
recapturing any increases in its value which have
been caused by the actions of society as a whole.

A third case for government intervention can be
made out in relation to the supply of 'public
goods'. If the benefits to be gained from a
particular use of land are great, but are impossible
to allocate accurately in the case of all those who
receive them, or if they accrue generally to most of
society, it may be easier for government to acquire
the necessary land and put it to the particular use,
in the name of that society. For example, the
defence of the realm or the provision of a strategic
timber reserve in case of war are public goods which
no individual would normally find it worthwhile to
supply by himself, and for which it would be
difficult to levy a direct charge by market means,
despite the fact that almost all members of the
society want to enjoy the sense of security which
such provision gives. However, the government can
undertake such actions, and is justified in view of
the failure of the market, in acquiring land for
both military training and installations, and for
forestry. Similarly, where there is a general
demand to restrict the right of landowners to
develop their property or to change its use in areas
of great scenic beauty, it may be easier for the
government to compensate such owners for the
diminution of their property rights than for the
owners to collect a payment from all those who visit
the area and enjoy the unspoilt view, and to exclude
those who are unwilling to pay.

A further form of the public-goods problem
occurs where there are many owners who would all
benefit if only they could act together. If one
improves his property he may not get a full return
on it while neighbouring buildings remain in an
unimproved state, but if all owners can be persuaded
to cooperate the whole area may increase in value by
more than the sum of the expenditures on the
individual properties. Similarly, the community may
obtain social gains, such as the improvement of
transport facilities or the provision of open space,
where government compels and coordinates the
readjustment of property boundaries and the
redevelopment of land belonging to many owners,
which might otherwise be prevented by the opposition
of a minority. Furthermore, once governments have
become involved in the supply of public goods, they
may be justified in both limiting the area under

particular uses and determining the areal pattern of
those uses in cases where the free-market allocation
of land would otherwise cause the cost of those
public goods to be higher. For example, densities
of housing development may be raised in order to
shorten the length of the sewers required to serve
them, with the effect that the area of urban
development will be constrained.

Lastly, governments may override the decisions
of markets where these are based upon an
unacceptable distribution of income. For example,
the provision of a minimum standard of housing is
accepted in almost all developed countries, even
where this requires the purchase of land by local or
central authorities and the construction of
dwellings for which only part of the full cost is
charged to the tenants. Such developments are often
at lower densities, and therefore use more land,
than the market would allocate if the poor who live
in them were obliged to compete without assistance
for dwelling space.

Thus there are powerful reasons for supposing
that the long-term welfare of society requires,
amongst other things, communal action to regulate
the use of land. But Scott (1980, pp. 174-177) and
others see the situation very differently. They
argue that the chief concern of governments is not
so much the general welfare as the need to protect
businesses and the owners of capital from any
breakdown of society which might be caused by the
conflicts which are engendered by market failure,
and which might inhibit the processes of
profit-making and capital accumulation. They base
this view upon studies of capitalist economies, and
especially those in North America and Western
Europe, but, as we shall see in Chapters 3 and 7,
there is no reason to suppose that land-use patterns
in centrally-planned economies are free from market
failure either.

CONCLUSION

Thus, there are a variety of circumstances
which may justify the regulation of land use by
governments on behalf of the community. Even if
neo-Malthusian crises in developing countries were
to place no more than moral obligations on the
richer economies to preserve their ability to supply
food, minerals or other land-based resources - and
it is unlikely that they would not impose some
adverse economic effects as well - there are likely

to be powerful reasons for government intervention arising out of market failure within those economies. However, there is no clear indication as to the extent to which the power of the state should be employed, nor as to the methods which should be used. For instance, in order to achieve a desired level of domestic food production it may be sufficient for the prices of farm goods to be raised by the use of import tariffs. Alternatively, deficiency payments may be made to farmers while trade remains unhindered; or, thirdly, government may intervene in the market to buy up products, and thus force up their prices. However, if it cannot be assumed that such incentives will call forth a sufficient increase in the ingenuity with which a declining area of farmland is used; or, if it is accepted that there are limits to the extent to which output per unit area can rise, as Malthus and some of his followers have suggested, other methods for protecting the potential for production would be required, and these might include the regulation of changes in the use of particular - and perhaps all - pieces of farmland in the country. Similarly, some people have argued that the protection of areas of scenic value will be achieved best by leaving them in the hands of farmers and estate owners who, they claim, have an instinctive feel for the countryside and a sense of vocation in the preservation of its character. Others, in contrast, believe that the basic motive of landowners is not preservation, but profit, and so have argued that only detailed land-use control will preserve such landscapes. We shall see what balance different societies have struck between state involvement and the market, and what choice of methods they have made to deal with the land problem in its many guises, in Chapters 4 to 7. All that it is necessary to note at this stage is that, since the late-nineteenth century, almost all of the nations with developed economies have come to accept that significant state intervention in land-use matters is justified.

However, by adopting extra-market means to however small a degree - and in some countries almost all land-use decisions are now taken by government - to allocate land among competing uses, societies have altered significantly the number and influence of the interested parties who may affect these decisions. Whereas only a few potential users may be in effective, direct competition for a site in a free market, all those groups and individuals who have any sort of interest may also press their views when the responsibility for allocating sites

among alternative uses rests with elected bodies and
their advisers. Therefore, in order to understand
the way in which the land problem is handled, it is
necessary to know something of the character and
strengths of the various types of actors who are
involved, and it is to this that we must now
progress.

Chapter 3

ACTORS AND OPTIMA IN THE LAND-USE DEBATE

"Land-use planning is not a science. There are
few, if any, formal guidelines or principles to
justify land-use allocations. There is no way
to prove that one city has too much land set
aside for industry or that another has not
retained enough parkland. Land-use decisions
are, for the most part, political ones.
Competing interests can be counted on to
advocate land-use choices that tend to favour
their positions."
(L. Susskind [1981], Citizen Participation and
Consensus Building in Land Use Planning, in de
Neufville, J.I., The Land Use Policy Debate in
the United States, Plenum Press, New York, pp.
183-4.)

INTRODUCTION

        In the previous chapter several reasons why
governments might feel the need to intervene in the
allocation of land amongst competing uses were
noted.   We saw that, although Malthusian and
neo-Malthusian threats are unlikely to affect the
developed economies directly in the foreseeable
future, shortages of food, fuel or minerals in some
parts of the world might lead to substantial falls
in the quantity which would be available generally
for import, or to unacceptably large increases in
their prices; and that, as a result of the
recognition of this risk, many developed countries
have felt obliged to protect their stocks of
agricultural, forest and mineral land. Moreover, we
noted that it is in the nature of the developed
economy to give rise to a wide variety of other
types of market failure, and especially those

associated with spillovers; and that therefore demands for the imposition of some form of land-use control are extremely likely, However, it was pointed out that the consequence of any substitution of planning for the market allocation of land is usually a considerable increase in the number of groups and individuals who are able to press their point of view upon those who make the decisions.

Not surprisingly, the large and growing number of participants, or 'actors', and the complex inter-relations between them, have attracted considerable attention as the role of government in land-use decisions has grown (Cherry 1982, Goldsmith 1983, Ley and Mercer 1980, Roweis and Scott 1978, Simmie 1974). However, actors often fall into a rather limited number of broad categories, each with characteristic sets of beliefs as to how land resources should best be used and relations with the other participants; and, for the purpose of this study, the following categories have been identified - the landowners or their tenants; pressure groups, many of which do not own the land involved in any decision; the professional planners and other advisers who are employed by government to assist in the making of decisions; and government itself, which adjudicates between the conflicting opinions of the others. Each category will be described briefly in this chapter in terms of the general pattern of land use which its members consider to be desirable, and the influence which they can usually bring to bear upon decisions; and generalised and simplified descriptions of some of the relations between them in free-market and centrally-planned economies are given in Figures 3.1 and 3.2 respectively. But, because the power of each type of actor varies widely between countries, we shall delay a more detailed discussion of the situation in the countries which have been chosen for individual study until later chapters.

THE LANDOWNER

The prime mover in any decision about the use of land will often be the owner or his tenant or leaseholder. However, his identity and his rights may vary widely between countries. In market economies he will usually be a single person, a household or a firm, although central and local government may also own large areas of land. But in the USSR and almost all of Eastern Europe a very large proportion of the land is in the hands of

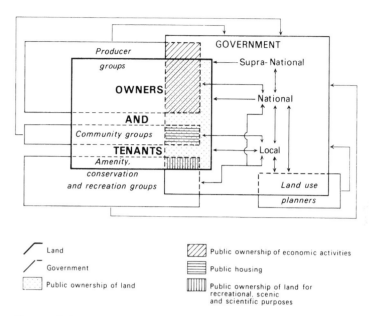

Figure 3.1:   Some Actors in Land-Use Decisions in Market Economies

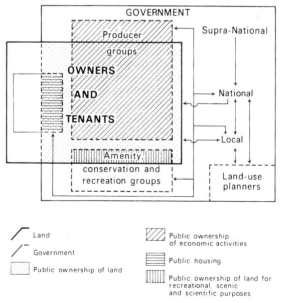

Figure 3.2:   Some Actors in Land-Use Decisions in Centrally-Planned Economies

agricultural cooperatives and the state. Similarly,
whereas in some countries ownership carries with it
the right to any game, timber or water on a site,
and to any minerals which may be found beneath it,
these are the subject of separate ownership in
others.   Thus, deposits of coal in the United
Kingdom belong to the state, but sand, gravel and
various other minerals do not; and in the United
States, where some mineral rights can be bought and
sold independently of the pattern of surface
ownership, disputes arise over the exercise of
ownership rights where the surface and sub-surface
uses are incompatible. However, for the purposes of
this discussion, it will be assumed that the
landowner is a private person or a firm, and that
owners have full rights to the use of the site,
untramelled by disputes with other owners of the
same land.
     According to the theory of perfect competition,
owners will attempt to maximise the return from
their land, and will alter the intensity and nature
of its use in accordance with circumstance.   Indeed,
they may be obliged to change the way in which it is
used if they are to continue to own or occupy it in
competitive situations;   and, as a result of this
process of adaptation, the aggregate rent obtained
from the land in any economy will tend toward the
maximum.   Among some of the most frequent types of
change in the pursuit of increased income in the
recent past have been the reclamation or improvement
of wetland for agriculture, the afforestation of
moorland, the building up of farmland, and the
intensification of city-centre land use by
skyscraper, office development. Some property
owners have also attempted to increase the value of
urban property by trying to persuade the planning
authorities to include it in conservation areas or
designate it as being of particular value (Dobby
1978, p. 65). In contrast, there have been other
changes which have involved decreasing intensities
of use, such as the abandonment of inner-city
property or the idling of urban-fringe farmland,
which may also have been motivated by the search for
profit, in the form of cost cutting.
     However, owners are rarely allowed to make such
changes with complete freedom; and the scope which
they are permitted will depend upon the concept of
ownership which any nation, or at least the most
powerful elements within it, have adopted. In most
developed countries there are constraints upon the
freedom of owners to inflict injurious spillovers

onto neighbouring properties; and many national and supra-national governments of market economies attempt to persuade owners to use land in desired ways through market management and the accompanying use of financial incentives. In communist countries, in contrast, it is considered to be inappropriate for the means of production, including land, to be under private control on any large scale, and so government is usually itself the major landowner. However, some land in both systems is in private ownership; and in most countries owners and users must apply to government for permission to make radical changes in the use to which it is put. These relations are represented in Figures 3.1 and 3.2 by an arrow from the owners to government, and by arrows back from the various tiers of government to the owners. That from the owners is to local government, because it is to that level of authority that application is normally made for permission for land-use change; but those from government to landowners exist at both local and national levels. At the supra-national level the European Economic Community (EEC) also attempts to affect some types of land-use decisions, but in Eastern Europe there is much less direct influence from COMECON for the wishes of the Soviet Union are usually passed down through the governments of Eastern Europe, and subsequently applied to individual landowners or users.

In spite of all this pressure some owners usually manage to obtain a greater degree of freedom of action with respect to land use than is generally permitted. For example, where a large proportion of the land within the boundaries of a planning authority is in the hands of a single owner, or where much of the local economy depends upon a single employer, the authority may feel obliged to follow his wishes in drawing up any zoning map or development plan, and in providing roads, drainage and other services. Thus, not only is the owner unlikely to be restricted in the use of his land, but also he may be able to insist on the adoption of plans which will enhance its value. Secondly, even where no single owner holds a dominant position in a community, firms or wealthy individuals may exercise undue influence through their ability to put pressure on voters who are their employees or tenants, and to buy favours by making contributions to the campaign funds of political parties or their candidates. Conversely, where the planning authority recognises that a grant of planning permission will give the owner or developer of land a substantial profit, it may be able to bargain with him in order to obtain some general benefit for the community, such as an urban park or the construction of a school, without cost to public funds.

However, it will be unlikely that, even amongst these powerful owners, the only aim will be the maximisation of returns. Some may be more concerned to obtain an increase in the capital value of their land, either at once or in the long term - that is, to use it as a hedge against inflation - than with the income it yields. Alternatively, they may be aiming to maximise the size of their holdings irrespectve of the rising marginal costs involved in further increases in the scale of their operations. Owners may be discouraged from developing their land to the extent permitted by the threat that neighbours will oppose their proposals, and so delay and make them more expensive; and the costs of redevelopment will limit the frequency with which owners will be able to adjust the use of land to changes in circumstance. Thus, even if all landowners are optimisers, and the pattern of land use is, in consequence, always coming into equilibrium with the rents which are available, it is highly improbable that the returns to all owners will ever be maximised simultaneously, if at all: some land will always be used inefficiently. However, many owners will not know what rents alternative uses would yield; and it is even more difficult to forecast accurately how rent levels might change in the future. Furthermore, even if all rent options are known, owners may be satisfied with their existing level of return.

Some writers have argued strongly that, in fact, if not in theory, owners in both market and centrally-planned economies have little external incentive for allowing their land to be put to the most profitable use in the short run, even if they are aware of the various incomes which alternative uses might yield; and they have gone on to claim that the widespread holding of land in the hope of long-run increases in its value has been the fundamental cause of the long cycles in the developed economies which were noted in Chapter 1, and in particular of the severe depressions which are a part of such cycles (Harrison 1983). As long ago as 1879 Henry George advocated the imposition of a regular tax on the value of land, net of any improvements upon it, in order to discourage such behaviour. Amongst the groups of owners who have been most widely criticised for such speculation in recent times have been the developers, and local and

national   government   acting   directly   or   through
government agencies. For example, it is frequently
alleged   that building companies in market economies
buy   farmland   close   to   growing   urban   areas   at
agricultural-use prices in the hope that eventually
it   will   be   zoned   for development,   and that,   in
consequence,   a large profit will be made when it is
sold   together   with   the   new   buildings   upon   it.
Meanwhile,   the   farming   infrastructure   may   be
neglected,   and the fertility allowed to decline to
the point where the land falls out of use altogether
for   a period. Similarly,   it is often claimed that
government bodies acquire land which they are not in
a   position to use,   but which they are unwilling to
release,   with the result   that   large   areas   from
which,   say,   slums have been cleared in the inner
cities in Britain, or sites which, for example, have
fallen   to   government   by tax default in the United
States, lie idle.
      But   government,   both local and central,   and
also   some other important types of landowner,   such
as   charities and trusts,   often,   by   their   very
nature,   have   aims   which   are very different from
those   which theory might ascribe to the   individual
or   firm.   The purpose of ownership for such bodies
is   less likely to be the maximisation of the return
in any narrow, financial sense, than the achievement
of   some cultural,   social or strategic aim   over   a
very   long   period   of   time.   For   instance,   the
preservation of fine examples of planned parkland by
a   body   such   as   the   National Trust in England is
undertaken   not to make money but to ensure that the
public   will   have   an   opportunity of studying such
landscapes   at first hand   in   perpetuity.   To   the
extent   that   such   owners   are   able   to   subsidise
'inefficient' land uses from trust funds or taxation
they   will   be able to sustain patterns of use which
may be very different from those which a free market
would   have   produced,   and to do so over very long
periods.   Nevertheless,   in   most   of   these
non-economic   cases of landownership there will be a
limit   of   subsidy   beyond   which   it   will   become
unacceptable   to   continue   the   use unchanged;   and
therefore   the pattern of potential rents from   land
is   always likely to find some reflection in that of
land use.

PRESSURE GROUPS

      A   second   category   of   participants   in   the
land-use   debate   is   that   of   the   pressure group.

These organisations vary widely in their interests,
size, resources, influence and methods, but are
almost always minorities within the population.
Some are composed of landowners, but many own little
or no land, although they often have clear views as
to how it should be used. Indeed, landownership is
a less important criterion for subdividing the very
large number of groups which exist than their
general aims. Those groups which are made up of
producers are usually concerned to protect, and if
possible, enhance, the incomes of their members,
while those which advocate conservation and amenity
generally find themselves in opposition to the
producers. The sporting and recreational lobbies,
in contrast, want to have access to the countryside
for the pursuit of a wide variety of activities, and
are frequently in conflict with both the
conservationists and the producers; and, last, there
are community groups, which are usually more
concerned with the quality of housing and the urban
environment than with rural matters. In general,
pressure groups are common in market economies,
where they usually, but not always, are allowed to
organise fairly freely, and to urge their opinions
upon government directly, as indicated by the arrows
in Figure 3.1 from 'producer', 'community' and
'amenity' groups to 'government'. However, such
organisations are rare and short-lived in the
centrally-planned economies of Eastern Europe, where
they are seen as a contradiction of the concept of
the leading role of the communist party in the
organisation of society. One independent
organisation which threatened to exert a major
influence on land matters in Poland - Rural
Solidarity - was suppressed by the government at the
first opportunity.

   One of the most successful of the producers'
groups in developed, market economies would appear
to be that of the farmers, for many governments of
such countries have been persuaded not only to
protect farmers from the fluctuations in the price
of their products, which arise from the vagaries of
weather, but also to guarantee a level of return
above that which the market would set. As a result,
the intensity of land use and the area of land in
farming are both greater than they would otherwise
be. Furthermore, farmers in some food-exporting
countries have obliged their governments to allow
sales to almost any country which is willing to buy,
even where that country is a potentially hostile
power, to subsidise exports, and to recompense them
when land which is surplus to requirements is taken

out of production for lack of such external markets. Also, several governments exempt farmers from some of the property taxes and development controls which apply to the rest of the economy; and, even in the USSR and Eastern Europe, collectivised farmers are allowed to retain small plots for the private production and sale of crops. This degree of influence is surprising for the typical agricultural sector in the developed economy is very small in both employment and the contribution which it makes to the Gross National Product. In food-deficit countries the success of the farm lobby is often based in part upon the argument that it is undesirable to become heavily dependent upon imports. Furthermore, agricultural communities are often over-represented on elected bodies, for declining rural population is not always accompanied by the redrawing of constituency boundaries. Nevertheless, there is widespread opposition from other sections of the population in many developed countries to the 'feather-bedding' of farmers; and, even in Eastern Europe, where food prices have been too low to stimulate sufficient output during much of the period since the Second World War, increases have provoked strikes and unrest.

Another producer group with a major interest in land use is that of the foresters. In several of the developed, market economies a large proportion of the forest land is in public ownership, but where there is also a substantial private sector, as in the United Kingdom and the United States, the commercial success of that sector is highly dependent upon the felling policy in the state forests and the system of taxation in the industry. Cutting and taxation regimes which seek to maximise government revenue in the short term will usually discourage private planting; and the foresters have usually sought, and often obtained, rates of felling which will not depress timber prices. Moreover, in some countries forestry interests have persuaded governments to adopt systems of capital taxation which allow payment at the time of felling rather than during growth, together with the opportunity to 'roll-over' receipts from timber sales into new planting programmes, and by so doing avoid tax. However, they have faced opposition from amenity pressure groups over the layout of plantations and the planting of open moorland in Britain, and from conservation groups in the United States over the clear-felling of mature forest, and especially of species such as the Californian Redwood.

A third important producer group is made up of the building trades and the mortgage and property companies. Their interest is chiefly in house building, and their lobbyists often argue that the supply of suitable land with permission for development is too small, that the tax incentives for home ownership are too low, and that government is not doing enough to promote the construction of public housing. Their hope is that large areas of greenfield land will be released for housing, industrial estates or shopping malls on the edge of the cities, and especially flat land with sufficient depth of soil to allow the easy construction of foundations and drainage systems. However, in this they frequently face opposition from amenity and conservation groups, which suspect developers of acquiring farmland at low prices, failing to maintain its fertility and agricultural infrastructure, and then arguing that, as it has become of little value for farming, it could be developed without a great loss of food-producing potential. They may also be opposed by governments which have adopted policies of protecting prime agricultural land; but they are likely to be supported by the farmers who stand to make substantial capital gains from such sales.

A fourth producer group is that associated with mining. Although the area of land which the industry uses is much smaller than that of either agriculture or forestry, the fundamental changes which opencast working in particular makes to the landscape, and the failure of the industry to restore many sites after use in the past, have aroused strong opposition from amenity and conservation interests. Moreover, because minerals only occur in rocks of a limited range of age, it is inevitable that their exploitation can only occur in a few regions. Thus, mining of many metalliferous ores can only take place in areas of hard, ancient, or of intrusive, igneous rocks, and these tend to be in mountainous districts, often of high scenic quality. Conversely, sand and gravel is largely found in lowland areas, but it is often overlain by well-drained soils of high agricultural capability.

Thus the producer lobbies often want to change the pattern of land use, and have considerable resources of their own with which to back their campaigns. Conservation and amenity groups, in contrast, tend to oppose change. They are usually voluntary organisations or charitable trusts, and they are often obliged to rely on public donations for support. As a result, their degree of success

in influencing government policy or individual
land-use decisions provides some indication of the
general attitude and strength of feeling of
society - or at least its more energetic and
articulate elements - towards land. At one extreme
are those groups which are generally middle-class,
conformist, and well-connected with the major
political parties. Such groups often have long
histories, considerable resources, and are able to
acquire or maintain land and buildings which they
wish to preserve. Furthermore, they may also
exercise influence over the areas adjacent to their
holdings by setting an example to neighbouring
owners (Tunbridge 1981, p. 104). The National
Trust in England and the Audubon Society in the
United States are important examplars for such
groups. Others, in contrast, are revolutionary,
separatist or utopian in their aims and methods, can
only afford to mount protests, and attract little
mainstream-political support. Many grew up rapidly
in the late 1960s and early 1970s in response to the
surge of concern over pollution and the supply of
food and oil, and have since withered away.

Some sporting and recreational groups are also
influential in matters of land use. Although they
usually do not own land at all, bodies such as the
Ramblers Association in Britain or the National
Rifle Association in the United States have very
large memberships, and have succeeded in opening
both private and public lands to a wide range of
recreational activities. Many of these activities,
such as walking, hunting, water-skiing and
snow-mobiling require large areas of land, and make
use of it in manners which are anathema to
conservationists intent on the preservation of the
natural flora and fauna. Furthermore, advocacy of
the rights of the public to visit mountain, forest
and lake, if need be in large numbers, for
recreation brings sporting groups into conflict with
those interested in amenity, which wish to preserve
the peace and solitude of such areas.

A separate category of voluntary bodies is that
of the community groups, which are concerned largely
with the welfare of the poor, and especially those
in cities. These groups are often radical in aim,
but poor in resources. Many have appeared since the
mid-1960s to campaign for the rehabilitation of
housing in inner cities, rather than its demolition,
to oppose motorway construction through
densely-populated areas, and to argue for the
provision of play areas for children. The
participants often live in rented accommodation,

including public housing. Such groups appear to have succeeded in obtaining recognition of their problems from government in recent years, but, to the extent that the urban environment is still characterised by wide contrasts in the density of building and provision of amenities, their demands for equal access to living space and land have yet to be met.

THE PROFESSIONAL ADVISERS

Before any decisions are made either about land-use policy in general or in response to particular proposals for land-use change it is usual, but by no means inevitable, for governments to seek the opinion of professional advisers. These advisers may be divided into two groups. Firstly and most obviously there are the land-use planners, but there are also a wide range of other types of adviser who assist government in connection with a variety of topics which may be related to land.

Many of the land-use planners in market economies operate in professions, whether they work in private firms or are employed directly by the authorities. For example, almost all such planners in the United Kingdom belong to the Royal Town Planning Institute - a body which acts not only as a setter of standards for entry into the profession but also as a pressure group which offers its views to government. Most planners in market economies are employed either directly or as consultants by the local authorities to draw up the fundamental documents of land-use regulation - the development or zoning plan - and to advise on applications for permission for land-use change (Figure 3.1). In centrally-planned economies, in contrast, where economic, but not necessarily land-use, planners are employed by definition, they usually operate only within the government machine (Figure 3.2), although specialist information has been sought by governments from the Academies of Science in all the Eastern European countries, either on particular topics or through the establishment of standing commissions. Thus, in both types of developed economy planners are in a relatively strong position to influence the decisions which governments make about land use.

Knox and Cullen (1981, pp. 885-898) have shown that the average local-authority, land-use planner in Britain believes strongly that the advice which he gives to councils is independent, and that he fulfils an important role as a balancer of conflicting interests, as a manager of the urban environment, and as a protector of the natural environment. However, planners are not necessarily disinterested in the advice which they give; nor do they always act in a purely professional manner. For instance, they are generally in favour of planning, which means that they have a bias against allowing the market to determine the use of land. Secondly, they have a professional, and possibly personal, career interest in seeing the adoption and implementation of the plan which they produce. Moreover, they are in a strong position to achieve this aim, for in most cases they control the information and the options which are sent forward to elected bodies, and, although "planning is not a science", it is possible to present arguments in such a technical manner that elected representatives, who are likely to be laymen in the matter, will feel unable to dispute the conclusions which have been reached (Kirk 1980, pp. 34-6). However, it should be remembered that the advice which is given is usually constrained by a general set of policy aims which have been chosen in advance by the government, and that therefore planners' scope for suggesting radical alternatives is limited. This is even more likely to be the case where the officials are also elected or change with the political complexion of the government, for in such circumstances they are likely to be little more than the creatures of the politicians. Thus a link exists in both directions, as indicated by the double-headed arrows in both Figures 3.1 and 3.2 between the national and local tiers of 'government' and 'land-use planners'. Nor may planners be as independent of producers as they appear. "Plans must conform to market principles to be capable of achievement" (Kirk 1980, p. 37); and, even in centrally-planned economies, the ability and willingness of the industrial ministries within the government to meet the targets of production which are set probably plays an important part in plan construction and the consequent pattern of land use.

    In addition to the planners other teams of specialists in agencies, commissions and ministries also advise government, especially at national level, on the formulation of land-use policy in both market and centrally-planned economies. To a large extent these bodies are the institutional counterparts of the actors who have already been identified, so that ministries of agriculture, for

example,   tend   to   argue for financial support for
farmers, while departments of mining resist attempts
to   legislate   high   standards   of reclamation,   and
environmental   agencies urge   the   strengthening   of
protection   for   areas   of   scenic   or   scientific
interest.   Some,   such as the Forestry Commission in
Britain, the Bureau of Land Management in the United
States,   and the industrial ministries   in   Eastern
Europe,   hold land on a large scale, whereas others,
such   as the Countryside Commissions in   Britain   or
the American Environmental Protection Agency, merely
advise about land owned by others.   Those which hold
land   are   shown   by the shaded areas in Figures 3.1
and   3.2.   In the market economies pressure groups
present   their   cases to these branches of the civil
service   or to their ministers,   as well as directly
to   members   of   the legislatures,   in the hope that
they   will not only accept them   but   advocate   them
strongly   and   rebut   opposition   to   them from such
other   branches of government as Treasuries,   which
often   wish   to   reduce   government   spending,   and
Ministries   of Labour,   which may want to   encourage
economic activity in order to boost employment;   and
it   has been suggested that   some   branches   of   the
civil   service   have   become   dominated   to   such an
extent   that   the   advice   which   is   offered   to
government   is   usually   favourable   to such groups.
For   example,   it   has   been   revealled   that
representatives   of   agriculture have come to occupy
an increasing number of the places on the Boards and
committees   of   the   National   Park   Authorities   in
Britain   since 1979 (Brotherton 1983);   and   it   has
been   alleged   frequently that the Forest Service of
the   United States has been 'captured' by the timber
producers   (Cullane   1981,   pp.   16-20).   If this is
true the position is serious for the advice which is
tendered   to   government   by   such   agencies   and
ministries   usually   forms   the   basis   not   only   of
policy,   but also of the legislative instructions, in
the form of laws and orders,   which are sent back to
these   same   agencies   and ministries for execution.
In centrally-planned economies,   in contrast,   where
independent   pressure   groups   do not usually exist,
differences of view are generated and debated within
the   relative   secrecy   of   the   government   machine.
Nevertheless,   in both systems influence is exerted
upwards   by   civil   servants   and   local   government
officers,   and it is often to the same people   that
the   execution of the decisions which have been made
on   the basis of that advice is entrusted,   and   who
are   thus   given   a   further   opportunity   to   steer
decisions   according to their   wishes.   Thus,   the

arrows in Figures 3.1 and 3.2 between the advisers
and government are double headed.

GOVERNMENT

But it is to government that the arguments and
views of all the other actors are eventually
expressed, either directly or through the medium of
civil servants and local government officials, with
regard to those decisions about the use of land
which can no longer be left to the workings of the
market. And it is from government, in the form of
elected representatives or those who have seized
power, that decisions are expected which, for the
reasons which were set out in Chapter 2, should be
in accord with the public and long-run, rather than
the individual and short-term, interest. That this
expectation is so frequently unfulfilled, not least
in the view of many of the other actors, is merely
to draw attention to the fact that there are at
least four reasons why any government may be no more
independent than any of the other participants in
the land-use debate, whether that government be
selected in the manner of the western or the
people's democracies, or by other means.
Firstly, the usual nature of representative
government is not conducive to the disinterested and
even-handed resoution of land-use conflicts. For
instance, the winners of elections are very likely
to be of the same opinion as their supporters, and
to favour their interests as a matter of policy.
Thus, Labour governments in Britain and Democratic
Presidents of the United States have both tended to
make more money available to help the redevelopment
of the central cities than their opponents.
Secondly, because elections are only held
periodically, and are not referenda on particular
issues, land-use matters are rarely commented on
directly by the electorate, with the result that the
decision-making bodies enjoy considerable freedom to
choose as they, or the majority party on them, or
the majority's leaders, wish. Third, although lower
levels of government are often called upon to advise
higher ones in questions of land use, conflicts may
arise between these levels in which local wishes are
overridden. For instance, national governments
which are concerned to protect the natural
environment may be opposed by local councils which
wish to attract new industry; or those who wish to
restrain expenditure may be obliged to subsidise
farmers to a greater degree than they want by such

supra-national   organisations   as   the   European
Economic Community. In the same way the General
Agreement on Tariffs and Trade (GATT) has power to
prevent the dumping of agricultural surpluses on
world markets at subsidised prices, and thus thwart
national policies; and, in the case of Eastern
Europe, it is arguable that the agricultural and
land-ownership policies which have been implemented
since the late 1940s, largely as a result of the
influence of the Soviet Union, are contrary to the
wishes of the rural population. Indeed, although
advice may flow in both directions between the tiers
of government in centrally-planned economies,
superior levels are almost always dominant and can
command the conformity of inferior ones (Figure
3.2).   Lastly, the intervention of government does
not necessarily eliminate the problem of market
failure.   Indeed, it may raise it to a larger
scale. For instance, the attraction of new industry
to an area may raise the local tax base;   but the
growth of population which arises out of such
development may spill over into neighbouring
communities,   and impose extra costs of housing,
schools and other infrastructural services upon them
for which there is little new revenue. Similarly,
city councils in both Britain (Self 1982, pp. 41-2)
and the United States have experienced difficulty in
persuading suburban authorities to accept some of
their inhabitants, and so assist in the process of
decongesting central urban areas;   and, at the
international scale, the growth of electricity
generation in Britain has led to complaints from
Scandinavia about high-level atmospheric pollution,
acidic rainfall, and the impoverishment of farmland
and lakes (Royal Society 1980, pp. 21-2).
    On the other hand, it should be noted that the
power of government to regulate land use is not
unlimited.   In many of the developed countries most
of the land is owned by private individuals and
firms, and so is much of manufacturing, the service
industries and housing. Thus governments in these
countries are obliged to rely largely on negative
controls over land-use change, or the persuasion of
owners to change their activities through the offer
of financial incentives. Opportunities for positive
planning exist only to the extent that land and
production are in public ownership, as is generally
the case in Eastern Europe. Secondly, for policy to
be effective it must be widely accepted or
powerfully imposed, and continuous.   Where a
consensus exists as to how land should be managed,
as has been claimed in the case of agriculture in

Japan (Ogura 1979, p. 607), it is possible for government to pursue a consistent policy; but where, in contrast, there is deep division between groups, as in the case of the treatment of development gains on land, or 'betterment', arising out of grants of planning permission in Britain, or where variances or rezonings of land can be obtained, landowners are likely to behave just as if there is a free market in land, and little will be achieved (Scott 1980, pp. 63-64). Moreover, in countries where government has imposed policies which are strongly opposed by large sections of the population there may be both evasion by the public and violent opposition leading to sudden policy shifts, such as those which have occurred in Poland since the Second World War over the pricing of agricultural products; and, in the extreme, governments may be obliged to abandon land-use regulation when the overwhelming need is for economic growth, rather than for equity.

CONCLUSION

        Several types of actors have been discussed in this chapter, each of whom has different interests to protect and optimal patterns of land use towards which he may wish to progress. However, it should not be concluded that every type of actor is involved in all land-use decisions, nor that all the interactions which have been sketched here occur in all developed economies. Different societies take decisions in different ways; and the scope for each type of participant varies accordingly. For instance, it has been argued that, while conservation and amenity groups are strong in Anglo-America and Western Europe, and community groups rather weak, real power lies with capital, in the form of landowners and producers, and with the bureaucracy, in the form of the planners (Kirk 1980, pp. 38-57, 60-66, 109-127). In Eastern Europe and the USSR, in contrast, truly independent or voluntary sectors are almost absent, for the only groups which survive for any length of time are usually those which have been established by the state. Furthermore, governments in those countries do not have to defend their land-use proposals at public enquiries, and there is no effective opposition to them from the general public (Shaw 1981, p. 301). The power of the political elite in the centrally-planned economy, in the form of its direct control over not only land and the other

means of production, but also over more junior
members of government and the administrative
machine, is very great. Indeed, these countries are
sometimes called 'command economies'. However, the
secrecy with which decisions are taken, and the
overlap which exists between the representatives of
industry, the planning agencies, government and the
communist party, make it difficult to assess the
degree of influence which each exerts.

> "Even at the level of relatively minor
> decision-making concerning the spatial
> organisation of the economy it is often
> difficult to determine the formal divisions of
> responsibility between different organs, while
> to fathom who actually has control in any
> instance is well-nigh impossible" (Pallot and
> Shaw 1981, p. 32).

Thus there is a major contrast between the mixed,
pluralist economies of Anglo-America, Japan, Western
Europe and some other parts of the world, and the
centrally-planned economies of Eastern Europe; and
the situation in the developed economies under
right-wing dictatorial and military governments in
Asia and South America which were listed in Table
1.1 may be different again.

Thus the opportunity which regulation affords
for participation in decisions about land use varies
widely. In fact, it varies so widely that there is
dispute as to where exactly power lies in some
countries, and as to what their political structures
might most accurately be called. It has been
suggested above that, at least when compared with
the countries of Eastern Europe, Britain, Japan and
the United States might be described as 'pluralist',
but it should not be forgotten that there are many
other models which have been used to characterise
them, including 'bureaucratic' (Kirk 1980, pp.
66-73), 'corporatist', 'majoritarian representative
leadership' and 'modern consumer populism' (Self
1975, p. 105). However, it is not the purpose here
to allocate any country to one or other of these
categories, but to draw attention to the differing
balances of power within them, as indicated by this
range of models, and to note that we shall return in
subsequent chapters to the description of some of
the actors who have been mentioned as part of the
assessment of the conditions which surround the
land-use-control systems in the countries which have
been selected for individual study.

Chapter 4

LAND-USE REGULATION IN BRITAIN

"In Britain, to put it plainly, land tends to
be put to its most profitable use." (G. Kirk
[1980], Urban Planning in a Capitalist Society,
Croom Helm, London, p. 32.)

INTRODUCTION

So far we have examined the land problem as it
affects the developed economies in general from a
Malthusian or ecological viewpoint, from a
market-failure or economic viewpoint, and from a
decision-making or political viewpoint. Now it is
time to turn to a more detailed study of the problem
as it is perceived and tackled in the small number
of countries which were selected in Chapter 1,
beginning with Britain.

Kirk's claim about Britain is surprising. It
suggests that, in spite of a major effort by the
state to regulate land use, especially since the
Second World War, it is still the pattern of rent
which determines the overall allocation, the
intensity and the areal pattern of land use, and
that landowners and producers have managed to avoid
the restrictions which have been fashioned against
them. If that is the case it is a most serious
indictment of much work by government, both central
and local, which has been intended to solve some
widely-recognised forms of the land problem, for it
has been a commonplace of public debate that,
insofar as the country has a high density of
population - as we noted in Chapter 1 - its land
resources are subject to great pressure, and that
this pressure can only be channelled equitably by
public control. More particularly, it has been
generally accepted that the unfettered private

ownership of land is highly unlikely to lead to an adequate supply and protection of uses which yield little or no profit, such as working-class housing, public parks, nature reserves or scenic landscapes, to achieve a desirable segregation of nonconforming activities, or to secure an adequate, strategic reserve of land for agriculture and forestry for the nation. As a result, Parliament has enacted legislation which, in Britain, but not in Northern Ireland, appears to have put severe limits on the changes which owners can make in the use of their land without giving public notice and obtaining permission from government; and in this Britain has served as something of a model for some other countries. Yet Kirk suggests that all this has been in vain.

We shall examine this claim by describing in broad outline the supply of, and demand for, land in the country in the recent past, the attitude of society towards land ownership, and the character and influence of some of the pressure groups which are involved in land-use decisions, before turning to a consideration of the extent to which government has taken control of land-use change.

THE SUPPLY OF, AND DEMAND FOR, LAND

Geographers have traditionally divided the land resources of Britain into highland and lowland zones. Most of the better lands for agriculture, forestry and settlement are to be found in the lowland zone, whereas the highlands offer chiefly coal, water power and beautiful scenery. However, such a division is only a broad approximation, for there is much good arable land to the north of the Tees-Exe line, especially along the east coast of Scotland and on the Cheshire-Lancashire plain. Similarly, the lowland zone is not uniform, for the best arable soils in the east are complemented by poorer land in the more humid Midlands of England. An alternative division of the land resources is provided by height above sea level. Except in the north-west of Scotland it is generally the case that land below 250 metres has been improved for agriculture or is built up or wooded, while that at higher altitudes is used only for rough grazing, forestry, recreation or field sports, or is not used at all. About a third of the country falls into this upland category. A third, and more detailed classification of land, has been made by the Ministry of Agriculture, Fisheries and Food. In

1976 it published an assessment of the quality of
the agricultural land in Britain, which covered
about three-quarters of the total area, and divided
it into five grades or qualities. It showed that
less than two percent fell within the first of the
five grades employed, and was "land with very minor
or no physical limitations to agricultural use";
about half was in the first three grades, which
included all the land which did not have any severe
limitation; and one-third was considered to be in
the last, and to be of little agricultural value.
Almost all of this, the poorest land, lies in the
highland zone, and above 250 metres (Table 4.1).

Many demands have been made upon the country's
land in recent decades, and in particular upon the
agricultural area. Notwithstanding the small
increase in the population (Table 1.3), the built-up
area grew substantially both before and after the
Second World War, as suburban sprawl in the 1930s,
made up of local-authority and private housing, and
the post-war, slum-clearance programmes - which
moved several million people out of congested
inner-city environments to low-density housing
estates on city edges, to the twenty-eight New Towns
and to the much larger number of Expanded Towns -
took over much farmland. Moreover, the development
of industrial estates, the provision of schools and
spacious playing fields, and the motorway programme
also required large areas. In addition, Coleman
(1978) and Smith (1981) have drawn attention to the
large quantity of farmland in the urban fringe
which, although not formally transferred from
agricultural to urban use, has fallen idle or is in
such quasi-agricultural uses as 'horseyculture'.

Some assessment of the impact of these changes
may be obtained from the government's annual
agricultural census - the June Returns. Using this
source Best has calculated that in England and Wales
an average of 25,000 hectares passed from
agricultural to urban uses each year during the
1930s, and about 15,500 per annum between 1945 and
1975. Comparable figures for Scotland are only
available since 1960, but when combined with those
for the rest of Britain they indicate that a total
of about 17,000 hectares was transferred to urban
use annually between that year and 1975 (Best 1981,
p. 86). Some of the effects of these changes may
be seen in Table 4.2. Much of the land which has
been built on has been in the lowlands, but, because
the areas of fastest urban growth have not generally
been those of the best arable soils, only a small
proportion of that land has been of the better

Table 4.1: The Quality of Agricultural Land in Britain in
            1976

Land Grade	Percent of all agricultural land
1	1.9
2	10.2
3	36.1
4	16.4
5	35.4

Source: Agricultural Economic Development Council,
         Agriculture into the 1980s: Land Use, National
         Economic Development Office, London 1977.

Table 4.2: Land Use in Britain 1961-2001

(Thousands of hectares)	1961	1971	Estimate for 2001
Improved agricultural land	11,615	11,316	16,540
Rough grazing	7,041	6,425	
Woodland	1,685	1,853	2,540
Urban land	1,689	1,871	2,410
Total area	22,744	22,744	22,744

Source: Best, R.H., Land Use and Living Space, Methuen,
         London 1981, pp. 46, 103.

grades (Best 1981, pp. 118-133, 146-148). The much
higher proportion of the land in these grades which
has been transferred to urban use in Scotland
(Scottish Development Department 1981, p. 4) has
not been sufficient to alter the general picture for
the country as a whole. However, it should be noted
that these totals do not include all of the land to
which Coleman and Smith have drawn attention,
namely, idle land in the urban fringe and that in
'quasi-agriculture'.

However, any description of the conversion of
farmland to other uses in Britain in recent decades
must also make reference to the growth of forestry
for, great though the expansion of the built-up area
has been, it has been forestry - both in public and
private ownership - which has made the greatest
single demand upon rural land. In particular the
Forestry Commission, which was established in 1919,
has bought and planted large tracts with conifers in
order to provide a strategic reserve of timber.
However, in almost all cases foresters have only
been able to buy land at low prices, with the
consequence that most of it has been rough pasture
of very low agricultural capability in the uplands.
Nevertheless, an average of 25,500 hectares was
taken from agriculture each year for this purpose
between 1960 and 1975 (Best 1981, p. 86).

All told, farmland declined during the 1960s
and early 1970s by between 40,000 and 45,000
hectares each year. But how significant was this?
Firstly, these transfers only affected about a
quarter of one percent of the agricultural land
annually; and secondly, the rate of transfer in the
lowlands was slower than in the poorer uplands.
Furthermore, the transfer of the better grades fell
sharply in the late 1970s as the British economy
moved into a deep recession, and government policy
shifted away from slum clearance and the building of
city-edge and New Town housing estates to the
rehabilitation of existing property and the
redevelopment of vacant inner-city sites.
Nevertheless, there will not be sufficient space
within the urban areas to provide the 190,000 to
250,000 public and private-sector houses which,
according to government estimates, will be required
each year during the 1980s to replenish the existing
stock and cater for changing needs (Phillips 1982);
and Best (1981, pp. 100-101) has suggested that the
urban area will rise from just over eight percent of
the country in 1971 to 10.6 percent by the end of
the century (Table 4.2). Moreover, Forestry

Commission plans up to the year 2000 involve the
expansion of the wooded area to 11.2 percent of the
country, and even so Britain will still be less than
twenty percent self-sufficient in its supply of
timber. However, the transfer of a further half
million hectares of the uplands to forestry could be
accomplished without adversely affecting either
farming or water collection, and careful
interdigitation of farm and forest land could
actually lead to increases in farm output (Centre
for Agricultural Strategy 1980, pp. 180-200).
Using these and other estimates Best has calculated
that the agricultural area will continue to fall
from seventy-eight percent of the country in 1971 to
72.7 in 2001, but he suggests that this is unlikely
to produce conflicts of critical proportions "until
well into the next century" (1981, pp. 102-103).

In spite of this conclusion there have been
frequent expressions of concern about land-use
change, and especially about the erosion of
Britain's agricultural potential (Coleman 1976,
1978a, 1978b). In view of this it is ironic that
another of the controversial land-use changes during
the 1970s and early 1980s was the expansion of
improved agricultural land at the expense of
moorland and marsh. The ploughing and fencing of
parts of Exmoor National Park, the demands by
farmers to drain more effectively large parts of the
Somerset Levels and the Halvergate Marshes in
Norfolk, and the widespread complaints about farmers
removing hedgerows and ploughing out footpaths and
sites of archaeological interest, have all alerted
the public to a form of land-use change which has
not been of importance in Britain since the
nineteenth century, namely, the colonisation of
wastelands by farming, but which now seems to be one
which economic circumstances favour again.

ATTITUDES TO LAND

If controversy over improvements which are
designed to raise agricultural yields marks a new
stage in the developing attitude of the British
towards land, it is by no means the first, or even
the strongest, disagreement which has been
expressed. For much of its modern history British
society has given fairly strong support to the idea
that the landowner should be allowed to do what he
wishes with his property, and this right has been
strengthened by the high cost of litigation, which
has meant that the great majority of the public have

not been able to seek redress for any spillovers from which they have suffered. However, there have been several disputes in which the public have taken direct action against landowners in recent times, and have subsequently obtained legislation which has strengthened their position. For example, crofting farmers in the north of Scotland argued in the late-nineteenth century that land belonging to the large sporting estates should be made available at nominal rents and with security of tenure to the rest of the population in order to relieve rural overcrowding and stop emigration. Some land was forcibly occupied for a time by the crofters, and, although they were expelled, their action was followed by legislation which gave them exceptional privileges in relation to the owners, including rent control and hereditary tenure; and in the 1970s they obtained the right to acquire the freehold of their land for very small sums, irrespective of the owner's wishes. Moreover, the Highlands and Islands Development Board was given the power in the mid-1960s to purchase land compulsorily which it considered to be underused, thus threatening Highland estates which could be improved for agriculture but were in use only for deer stalking and grouse shooting. However, this power has never been used.

A second example of such action by the public was the mass trespass of walkers on Kinder Scout in the Peak District of Derbyshire before the Second World War. As in the Highlands, the owners' rights to protect their game-shooting land from disturbance by the public came into collision with the desire of the urban population to escape for recreation to the cleaner and contrasting environment of the moorlands; and, in spite of the fact that the trespassers were repulsed and efforts to legislate access to such areas in the 1930s failed, the National Parks and Access to the Countryside Act of 1949 marked the eventual acceptance of the idea that these areas should be made available to the public at large.

There has also been controversy in the more recent past over the question of 'betterment'. Under the Labour government of the late 1940s Parliament passed the Town and Country Planning Act of 1947, which not only required local authorities to zone all land for particular classes of use, but also nationalised all development gains to the value of land, and set up a system by which they were repaid to the state when land was developed. In doing this the government was acting on behalf of

tenants and workers in the cities, many of whom were
its supporters, who obtained no benefit from rising
land values, but who were adversely affected by the
effects of such increases on rents and the cost of
land acquisition by public authorities for housing,
open space and other public facilities. However,
when the Conservatives, who enjoyed the support of
many landowners and developers, were returned to
power in 1951 they abolished the charge on
betterment, and the increases in value reverted to
those owners who were fortunate enough to be granted
permission to intensify the use of their land in
some way. But, so deep was the dispute over
betterment that the process of legislation and
repeal was repeated twice again as the two political
parties alternated in government during the 1960s
and 1970s, culminating in the repeal of the last
Labour government's Community Land Act after the
return of a Conservative administration in 1979, and
a reduction in the Development Land Tax. However,
the Tax was not removed entirely, and so some of the
increase in land value arising out of the granting
of planning permission is still being captured for
the benefit of society as a whole, chiefly from the
major commercial developers. Thus there may have
been some convergence of opinion in the matter of
betterment over this period, and the saga of
legislative reversals may be at, or close to, its
end.

However, there were other, very live, issues to
do with land in Britain in the 1970s and early
1980s, one of which was the question of the
countryside. In a recent book Best has claimed that
British attitudes towards land are based upon a
series of potent and widely-believed "land-use
myths", amongst which is the belief that

> "Each year urban sprawl engulfs still greater
> amounts of good land so that, before long, most
> of our precious countryside will be completely
> submerged beneath bricks and concrete" (Best
> 1981, p. 186).

This description of British attitudes agrees with
that of Prince and Lowenthal (1965) who, in a
description of the attitudes of those who they
consider to be most active in creating English
landscape tastes, have identified, amongst others,
preferences for the bucolic, the picturesque and the
deciduous, all of which are permeated by a
strongly-developed antiquarianism. Thus John
Constable's view of the lowlands, with its emphasis

on the variety and balance between the open spaces of the fields and the sense of enclosure provided by the hedgerows and the trees within them; its suggestion of the idyllic nature of country life, provided by the gentle activity of the rustics who are portrayed; and the evocation of the timelessness of that life, by the inclusion of ivy-covered walls and the towers of ancient churches, is still the picture-postcard image which is in demand today. Similarly, the publicity of the tourist industry and the choices of pictures used for calendars suggest that Sir Edwin Landseer's and Sir Walter Scott's powerful encapsulations of the Highlands of Scotland as a bleak and rugged wilderness inhabited by plaid-wearing, pastoral clansmen has continued to dominate perceptions of what upland Britain should be. It is hardly surprising, therefore, that encroachment upon rural areas by urban development, the grubbing up of hedgerows in the lowlands, and the 'blanket' planting of conifers in the uplands, should all have been vigorously attacked during the present century, or that rapid changes in the landscapes of rural areas could have led to the publication of a book entitled <u>The Theft of the Countryside</u> (Shoard 1981). In other words, Cowper's claim that "God made the country, and man made the town" continues to find a powerful echo amongst a population which has been highly urbanised for more than a hundred years, and which values the contrasting environment of the countryside, and its access to it; and changes in rural landscapes which do not accord with the public's perceptions of what those landscapes should look like evoke strong and hostile reactions, and strengthen the demands for government intervention and control. However, it should be noted that these views are not always shared by farmers and foresters, and that the Conservative government was unwilling to impose any real control on the right of farmers to improve moorland in the <u>Wildlife and Countryside Act</u> of 1981 unless compensation was paid to them for the consequent loss of income.

Of course, it is not surprising that there should have been conflicts in relation to the matters referred to above, for in all of them the interests of landowners and of the public have been directly opposed. However, there has been a general tendency for the rights of owners to be reduced in the interests of other groups within the community, even if progress in that direction has not always been smooth or continuous, and as a result there are some matters today which have passed the stage of

controversy, and about which there is now a notable
consensus of opinion, with important consequences
for the way in which the use of land is controlled
in Britain. For example, it is now commonly
accepted that the unplanned urban growth of the
Industrial Revolution and the nineteenth century
created environmental problems which can and should
be avoided in future through the use of controls
over land by government; and there has been a
similar consensus of opinion in Parliament and
outside since the 1940s about the need to use such
controls to restrict the rate of conversion of
agricultural land to other uses, and to protect
areas of outstanding historic, scenic or scientific
interest. In particular, one of the most
widely-believed and exaggerated of Best's "land-use
myths" claims that

> "Agricultural output will be gravely
> endangered, and food shortages a serious
> possibility, if continuing losses of productive
> farmland to urban use and afforestation are not
> soon reduced substantially" (1981, p. 188).

This attitude may hark back to the time of Malthus,
but it is much more likely to be a legacy of the
food crisis and rationing of the Second World War,
when the German submarine blockade of Britain
rapidly exposed the strategic weakness involved in
dependence upon imports for three-fifths of the
country's temperate foodstuffs. However, those
conditions have disappeared long since for, by the
late 1970s, Britain was about seventy percent
self-sufficient in such foods. Furthermore, the
country had become a member of the EEC, in which one
of the chief problems was the disposal of surplus
farm products, rather than any further need to
protect agriculture's land base. Nevertheless,
expressions of concern from some eminent academics,
based upon the findings of the Second Land
Utilisation Survey of Britain in the 1960s, the
world cereal shortage of the early 1970s, and fears
about the long-term supply and price of
petroleum-based fertilisers and pesticides, were
reflected in the government's White Paper on
agriculture in 1975, which stated that

> "The government take the view that a continuing
> expansion of food production in Britain will be
> in the national interest... There are greater
> risks than in the past of wide fluctuations in
> price and world shortage [of food]... The

finite area of good quality land - declining
with the annual loss to other uses - imposes a
constraint on increased output" (HMSO 1975,
pp. 1, 8)

a view which was only slightly modified in the
subsequent White Paper of 1979 (HMSO 1979, p. 7).
    Thus British attitudes towards the ownership
and use of land have many facets. Some are based
upon class-consciousness and self-interest; others
are founded in received perceptions of what is
beautiful in landscape; and some are the consequence
of misconceptions about the supply of, and the
demand for, the land resources of the country. It
is not surprising that, backed by such a heady brew
of money, emotion and prejudice, land use, and the
best way to control it, have given rise to varied
and prolonged controversy.

SOME IMPORTANT ACTORS

    Debates about land in Britain have been
conducted by all the different types of actor who
were described in Chapter 3, and many of the general
characteristics of those actors were outlined
there. However, the identity and character of some
of the most influential of those in Britain, and
especially the most powerful of the producer and
amenity organisations, remain to be described.
    About half the land of Britain is owned by
private individuals. Members of the landed
aristocracy and gentry, each of whom own between
five hundred and several thousand hectares, together
account for about thirty percent of the whole; about
an eighth is held by central government and its
agencies, of which the chief is the Forestry
Commission; a twelfth is in the hands of local
government; about a fifteenth is owned both by
companies and by charities and trusts; and many
millions of house-owners and farmers, none of whom
own large areas, account for about a fifth of the
total (Massey and Catalano 1978, pp. 59, 94-97,
101). Of these groups one of the most interesting
is that of the large private landowners, while
another is made up of the property and insurance
companies and the pension funds. Both are strongly
opposed to the taxation of land values or of
betterment, and, in so far as most of their holdings
are rural, have a strong interest in the prosperity
of farming and forestry. It is hardly surprising
that they are hostile to the suggestion that

planning   permission   should   be   extended   to
agricultural land use. Their views are important to
land-use   decisions   in   Britain   because   they   are
strongly   represented   in   Parliament.   A third of the
hereditary holders of titles, who are thus members
of   the   House   of   Lords,   are   also major private
landowners (Perrott 1968, p. 149); many   of   the
large private and corporate land-holders support the
Conservative Party; and the register of interests of
Members of the House of Commons shows a

> "surprisingly   high   number   of   business
> connections   to   the   land/property   sector,
> through both property companies and financial
> institutions"   (Massey and Catalano 1978,   p.
> 169).

Amongst   the   producers,   the National Farmers'
Union and its Scottish equivalent are of particular
interest.   Since   the   war   they   have successfully
opposed   all   suggestions   that   land-use   controls
should be extended to agriculture;   and they were
able   to   persuade   the   government   to amend the
Wildlife   and   Countryside   Bill so as to grant
compensation to farmers who, because their land had
been designated for conservation purposes,   would be
unable to improve it.   They have   constantly   urged
government   to pay high,   guaranteed   prices   to
farmers - with considerable success in the cases of
cereal,   dairy,   hill and upland production - and to
exclude rival products from the British market, even
where they would have come from other members of the
EEC,   with whom trade is supposed to be unimpeded.
Agriculture   has   also   been   relieved   of   any
requirement   to pay local rates.   The reasons for
these successes are various.   The Unions claim a
large membership among farmers,   are well organised,
and regularly lobby ministers   and   combat public
criticism.   They have very close ties with another
powerful   pressure   group,   the Country Landowners'
Association; and both are closely connected with the
Conservative   Party.   Not   only   does   that   party
include   many   farmers   and   landowners   among   its
members,   but it also draws much support from rural
areas;   and it has been in power for two-thirds of
the time since the abolition of food rationing in
the   early 1950s.   Furthermore,   farmers   have   been
assisted by Britain's entry into the EEC in 1973,
for the agricultural interests on the continent have
been even more influential than those in Britain,
with the result that the level of agricultural
support has been higher than the British government

has wished. The consequences of the farmers' successes have been a higher level of agricultural output than the market would otherwise have justified (Bale and Lutz 1981, pp. 8-22), and this has resulted in a greater intensity of land use, the alteration of the landscape by the grubbing up of hedges and copses, and the reclamation of marginal areas, to the dismay of amenity and conservation groups.

Another of the producer groups which are frequently opposed by conservationists is what Ambrose and Colenutt (1975) have called "the property machine". This is composed of the Country Landowners' Association, the National Association of Property Owners, the large building companies, and the road-transport organisations. This lobby is generally in favour of the development of land; and it enjoys direct access to the land-planning activities of local government through its participation in the production of Joint Land Availability Plans with the local authorities, in which the need for building land in the near future in each area is set out. Amongst its members the large building companies are of particular interest, for they try to build up 'banks' of land in their ownership, or over which they hold an option to buy, if and when planning permission for development should be granted. This they do for several reasons, amongst which are the expectation that the increasing value of the land will more than offset the cost of holding it until it is developed, and the hope that land which has been acquired at agricultural-use value will jump in price once planning permission has been received for its development (Massey and Catalano 1978, pp. 109-113). Unfortunately, little is known about ownership, or options to buy, in advance of applications for planning permission, and so the exact extent of land 'banking' cannot be established. However, a survey of land with planning permission for private housing in England in 1975 revealed that by mid-1977 no work had begun on a third of the sites, and that the most usual reason for this was because the land was being held for speculative purposes, especially in metropolitan areas. On the other hand, these sites only accounted for about a sixth of the land which was included in the survey (Department of the Environment 1978). Nevertheless, as planning permissions lapse after five years if a start has not been made, these figures probably exclude large quantities of land which are being held by building

companies with a view to the long-run appreciation
of its capital value, for which planning permission
had not yet been sought. It should also be noted
that the property lobby, and especially the building
companies, were successful in persuading central
government to release extra supplies of agricultural
land for development in the early 1970s and again in
1983, and in so doing to override the views of local
governments and planners who wished to restrain
development.

Other producer groups include the large,
private mining companies, which have obtained
permission to explore and develop the mineral
resources of the National Parks, in spite of vocal
opposition from amenity groups, and the nationalised
industries and New Town Development Corporations,
which are exempt from some of the planning controls
which govern private developers. All of these
groups are involved with land-use and landscape
change on a large scale, and all enjoy much greater
resources than the average amenity or conservation
group.

However, what the amenity and conservation
groups lack in strength they make up for, at least
in part, in number. For example, Ratcliffe (1981,
p. 305) has reported that the number of local
amenity societies rose from about two hundred in
1957 to over 1,250 by the early 1980s, while Lowe
and Wibberley (1981, Appendix 2) identified
eighty-one national, voluntary groups involved in
rural conservation alone. One of the most famous is
the National Trust. Since its foundation in 1895
it, and its sister body in Scotland which was
established in 1931, have acquired more than 220,000
hectares of land, either by gift or covenant,
including seven hundred kilometres of coastline.
Their holdings amount to about one percent of the
country, and make the National Trust the largest
private landowner in Britain. Land which is held by
the Trusts is inalienable; and thus agricultural
estates, scenic uplands and undeveloped coasts in
their hands must be maintained in the form in which
they have been received, although it should be noted
that government has overridden inalienability on
occasion, with parliamentary approval, and permitted
development of Trust land where the national
interest has, in its view, been of paramount
importance. The National Trust campaigned in favour
of National Parks, and owns much land within them;
and it laid the foundation for the idea of the
Heritage Coasts in its Operation Neptune of the late
1960s, which was designed to bring as much as

possible of the unspoilt coastline under the Trust's control. In view of its remit it is perhaps not surprising that it generally gives greater emphasis to the conservation of the country's heritage, including whole estates and villages in its ownership, than to the immediate requirements of the people within them; and it is concerned to maintain the appearance of the landscape even if this leads, say, to the gentrification of restored urban property and neighbouring buildings. The strength of the Trusts, like that of the farmers' organisations, lies in the size of their combined membership, which is in excess of one million, and in the fact that they have "always had friends in high places", and especially among the landed classes (Tunbridge 1981, p. 115). As a result they were given the privilege of holding property in lieu of death duties in 1937, in return for guaranteeing public access to it, as well as being recognised as being charities for the purposes of taxation.

Other influential conservation bodies include the Council for the Protection of Rural England (formerly the Council for the Preservation of Rural England), the Georgian Society, and the Civic Trust. These organisations have been effective in persuading local authorities to protect both rural and urban areas which are of beauty or historic interest through the commissioning of reports of high quality (Sharpe 1975, p. 337) and by persistent lobbying; and the Civic Trust was largely responsible for the passage of the Civic Amenities Act of 1967, which gave local planning authorities the right to designate conservation areas in towns.

The growth of these and many other voluntary conservation organisations in Britain has several causes. Dobby (1978, p. 65) suggests that self-interest may have played a part, as owners have seen the increases in value which recognition by a planning authority can bring to their property, while Sharpe (1975, pp. 337-339) has claimed that there has been a collapse of public faith in the ability of the planners to protect landscapes which the public likes. Whatever the reason, articulate, middle-class opinion in Britain has moved away from support of farmers who have so altered the appearance of the countryside, from big developers who have replaced idiosyncratic, Victorian city centres with banal and windy concrete and glass, from industries which have threatened to despoil National Parks and coasts, and from local councils who, in addition to letting all the other changes occur, have torn down the inner cities and built

soulless housing estates on the urban fringes.
     Nevertheless, it is the producer groups which
appear to be favoured at present by the authorities
in Britain. For example, Darke (1979, pp. 14-17)
has shown that, in drawing up the South Yorkshire
Structure Plan, government invited eighty-eight
percent of the property, commercial and industrial
organisations which had made representations to
participate in the Examination in Public of the
draft plan, and eighty-one percent of the government
bodies, but only forty-two percent of the voluntary
organisations (defined in his study as the amenity
groups, leisure interests, organised labour and
parish councils), and a third of the individuals.
Secondly, in 1981 capital grants which were paid to
farmers in Britain by the government, and which have
financed in part the type of 'improvements' which
have been so widely criticised by conservationists,
were more than seven times as large as the total
budgets of both the government-funded and
independent conservation agencies; and the total aid
to agriculture was worth more than forty times as
much (Clayton 1982, p. 4).
     In addition to the independent organisations, a
number of official advisory bodies have been
established by central government, which are funded
out of taxation. The Nature Conservancy Council was
established in 1949 (as the Nature Conservancy) to
set up and maintain nature reserves, and to advise
government on nature conservation. More than 150
reserves and four thousand Sites of Special
Scientific Interest have been established by the
Council, covering about half of one percent of the
country, but it has identified more than four
percent as being worthy of protection (Nature
Conservancy Council 1977). Secondly, the
Countryside Commission, which has taken over the
work of the National Parks Commission, and the
Countryside Commission for Scotland, are charged
with keeping under review all matters to do with the
conservation of natural beauty in the countryside,
and the need to secure public access to it. The
chief impact of the Commissions upon land use has
been through their advice to national government
over the identification of areas worthy of
designation as National Parks or Areas of
Outstanding Natural Beauty, and to local government
in relation to applications for permission for
development in the countryside. However, neither
tier of government is obliged to accept the advice
which is tendered by any of these bodies; and it was
suggested, although subsequently denied, that

central government failed to reappoint the chairman of the Nature Conservancy Council in 1983 because the Council had been too diligent under his leadership in identifying farming areas requiring protection (The Times, 28 January 1983, p. 10).

CONTROLS OVER LAND-USE

Notwithstanding this apparent preference on the part of the authorities for producers rather than environmentalists, a substantial and varied body of controls by government over land use has been created in Britain. What is more, some of these are of ancient lineage. Government and Parliament have been involved with the layout and ownership of rural land at least since the time of the Enclosure Acts; and it could be claimed that the first attempt to control the pattern of urban development was taken by the Crown, although unsuccessfully, in London after the Great Fire of 1666. However, the modern, and continuous, development of controls probably dates from the Artisans' and Labourers' Act of 1868, which led to the control of the character of residential development by local government, though not its location, and gave local authorities the power to close and demolish insanitary or dangerous houses. Further legislation with the same aim of ensuring minimum standards of construction and facilities in 1909 also gave local councils the right to build houses; and these powers were strengthened again by the 1919 Town and Country Planning Act, the Housing Acts of the 1920s, and the Housing and Slum Clearance Act of 1930. Under the last of these more than 300,000 houses were demolished before the Second World War.

The 1919 Act also allowed local authorities to take positive steps to plan the future areal arrangement of development, and this was followed by the 1932 Town and Country Planning Act and the Restriction of Ribbon Development Act of 1935. However, the control over land-use change afforded by this legislation was weak. Participation in planning by the local councils was optional; the mechanisms for the approval and amendment of plans were cumbersome; and the threat of having to pay compensation to landowners and developers who were prevented from carrying out their proposals discouraged local authorities from taking decisive action to deflect development away from the sites preferred by the market. By 1942 only about five percent of England and Wales was covered by approved schemes, although plans were being prepared for about two-thirds of the country (Cullingworth 1979, p. 30). Thus, land-use change was subject to very little control before the Second World War; and in those cases in which government or its agents were empowered to take land into public ownership, such as for council housing or afforestation, they were obliged to pay the market price, and so the choice of sites was determined to a substantial degree, not by 'good planning reasons', but by market forces. On the other hand, important principles were established concerning the rights and duties of local authorities to carry out both slum clearance and new development on a large scale, and to control the use of land not in their ownership, albeit at the risk of having to compensate owners for losses of development rights.

After the Second World War, however, circumstances were very different. The bombing of London and other major cities not only tore great holes in the urban fabric, clearing away some of the nineteenth century's legacy of mean housing and insalubrious intermixture of industrial and residential development, but also was perceived to have provided a _tabula rasa_ on which a better environment could be constructed. It was widely felt that, in the second rebuilding of London, the opportunity to coordinate the requirements of users in a properly-planned development should not be lost, as it had been after the Great Fire. Furthermore, the suggestion of the 1944 White Paper on _The Control of Land Use_ (HMSO) that many elements of the general reconstruction programme were closely connected with the harmonisation of the competing claims on land had been generally accepted. Thirdly, the overwhelming victory of the Labour Party in the elections of 1945 brought to power a government committed to advancing the interests of the ordinary man who wanted a decent house in a clean environment, and access to the countryside for recreation. In such circumstances the interventionist recommendations of the Barlow, Scott and Uthwatt Reports found a ready acceptance, and a series of major pieces of legislation affecting land use was passed between 1945 and 1950.

Perhaps the most important of these was the 1947 _Town and Country Planning Act_. This imposed on local authorities for the first time the duty to draw up a Development Plan for the entire area of their jurisdictions, showing the use to which each parcel of land should be put. In addition, almost all development, but by no means all land-use change, became subject to planning permission; and permissions, which were granted by the same local authorities, had to have regard to the Development Plan. Local authorities were obliged to notify central government of any major development which they were proposing. However, development by central government bodies, the statutory undertakers such as British Rail, and by farmers for farming purposes, were outside the system of control, as were the agricultural improvement and afforestation of land. Nevertheless, almost all proposals for land-use change by private developers or public companies were brought into the control of the elected county and county-borough councils. Furthermore, planning authorities were relieved of the compensation problem. The right of owners to develop land, and the increase in land values arising out of the granting of planning permission, were both nationalised; and developers were obliged to return the full amount of the increase, or betterment, to the government in the form of a Development Charge. Thus weaknesses of earlier legislation were removed, and a nationwide system of land-use control was established.

Within this system both central and local government were able to exert considerable influence upon the pattern of land-use, especially in the urban areas, but it was central government which held the greater power. For instance, although local authorities held the initiative in drawing up the detailed form of the Development Plans, it was central government which laid down the broad policies, such as the general presumptions against the development of prime agricultural land and sporadic development in rural areas, upon which the Development Plans were based. Moreover, all Development Plans had to be submitted to central government for approval, and in many cases ministers took the opportunity to modify them and bring them into line with central-government policies. Any permission for development which was a major departure from the Plan also had to be notified to central government, and there was a general power to call in any proposals for decision which seemed to central government to raise issues of national importance. All appeals against refusals by planning authorities to permit development, or against the conditions imposed upon any permission, were also decided at the centre; and proposals for certain types of development, including opencast coal mining, were dealt with entirely at the national level. Thus, the opportunities for central government to influence the use of land were very considerable, and they have been extended since the Act was passed, chiefly by a widening of the range of development proposals which must be notified to the centre by the local authorities.

The results of the Act were considerable. The pre-war practice of ribbon development was halted, and the much older problem of the intermixture of nonconforming land uses within built-up areas was gradually reduced. Of course, the legacy of the pre-control era did not disappear overnight, for existing uses were allowed to continue, but eventually redevelopment removed most of them, and the legislation prevented the reappearance of the problem in the same form elsewhere to a large extent. Moreover, the Act was modified in 1955 to include the power to designate Green Belts around provincial cities; and by the late 1960s nine percent of the country was subject to the extra control against development which these Belts provided. Also many more types of land-use change by central-government bodies and the statutory undertakers have been added to those which must be notified to local authorities; and, in many of these, applications for planning permission can now be refused by local councils without liability to pay compensation. However, in some other respects the Act did not work well. For example, the process of drawing up and regularly reviewing the Development Plans proved to be much slower than had been expected, with the result that they could not be adapted quickly to changes in, say, the growth of population or road traffic, and rapidly fell out of date. And so, before the last of the original Plans had even been drawn up, and before many had been subject to the first of what should have been quinquennial reviews, the Town and Country Planning Acts of 1968 (for England and Wales) and 1969 (for Scotland) made radical changes to the system.

Under the new legislation, which eventually came into effect after the reorganisation of local government in 1974 (in England and Wales) and 1975 (in Scotland), a distinction was drawn between strategic and local planning. Counties, the Greater London Council and the Scottish Regional Councils became responsible for the preparation of strategic or Structure Plans, which must be submitted to central government for approval, if necessary after modification, while the second tier of local authorities - the Districts and London boroughs - draw up Local Plans, which should be related to the Structure Plans, and must be approved by the strategic-planning authority. Structure Plans are very different from the former Development Plans. In the first place they include, and consist primarily, of a statement of the general planning policies of the local authority. Central government has advised (Department of the Environment 1974) that these statements should indicate how national trends in the economy and in society will affect the requirements for land for both industry and housing, and perhaps also for the conservation of the natural environment, the provision of recreation and tourism, the location of shopping centres, and land reclamation, over a period of fifteen years. However, it is no part of the Plan to show exact sites at which any of these developments are to take place, except where very large industrial developments are to be accommodated; and a detailed map is not required. Local Plans, in contrast, are much closer to the old Development Plans, showing in map form how the different demands for land within the District will be balanced and distributed. District councils are the authorities of initial receipt for most types of development proposals, but there is a 'call-in' procedure in order that both the strategic-planning authorities and central government may take over the decision on any proposal which they consider to be of regional or national importance.

Thus a hierarchical structure of land-use planning and development control has been built up over a period of fifty years in Britain; and, as the degree and extent of control has increased, so have the size and resources of the local government planning authorities. Early attempts at land-use control before the Second World War were the responsibility of about 1,400 authorities in England, many of which were small and had very limited financial resources and few professional staff. Provision was made in the 1919 <u>Act</u> for joint action by councils in order to mitigate these problems, but few joint committees were estalished, and the system of control remained very weak. Under the legislation of 1947, in contrast, responsibility for planning was concentrated in the counties and county boroughs of England, of which there were only 124. These were much larger organisations, which could afford and justify the appointment of suitable teams of professional planners. More recently, the 1968 <u>Act</u>, which recognised the need for land-use control at the city-region scale, was passed in the expectation that the Royal Commissions on Local Government, which were then sitting, would recommend

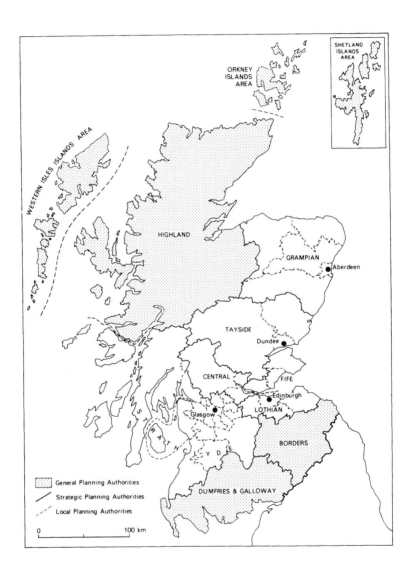

Figure 4.1: Strategic and Local Planning Authorities in
Scotland since 1975

a further reduction in the number of authorities, which they did. Thus, the number of strategic-planning authorities in England became forty-seven; but conversely, the number of local-planning authorities increased greatly to 363. Similar changes occurred in Wales and in Scotland, except that the number of local-planning authorities also fell north of the Border, and large areas of the northwest and south of the country with few inhabitants continued to have a single tier of control in the form of 'General Planning Authorities' (Figure 4.1).

It should also be remembered that local government has at least three other sets of duties and powers which concern the use of land. Under the housing legislation councils acquired and demolished slum property on a huge scale in most of the major cities between the mid-1950s and the mid-1970s, and have built new housing estates on an even larger scale to cater for the displaced population and new households. By 1980 almost a third of the housing stock was of local authority origin, most of which was at densities which were lower than that of earlier working-class accommodation, and on land which previously had been used for agriculture. However, after the passage of the <u>Housing Acts</u> in 1969 and 1974 the emphasis shifted to the improvement of older residential areas, rather than their demolition; and new, central-government machinery had to be set up to deal with the empty land left by the earlier slum-clearance programme in several of the largest cities, for it had proved to be beyond the ability of the local councils to deal with this on their own. Secondly, local authorities have the power to purchase and reclaim derelict land, and substantial grants have been made available to those in areas of high unemployment to do this. Some authorities have been very energetic in the matter, especially those in places such as Fife, Lancashire and West Yorkshire, which have suffered heavily from the legacies of coal mining, and where the local councils have wanted to clean up the environment in order to attract new industries, but the area involved has been small in relation to the total land resources of the country. Thirdly, the <u>Civic Amenities Act</u> of 1967 has strengthened the powers of local authorities to control development in areas of architectural or historic significance. Under the <u>Act</u> such areas may be designated Conservation Areas, as a result of which a much wider range of controls may be applied than normal

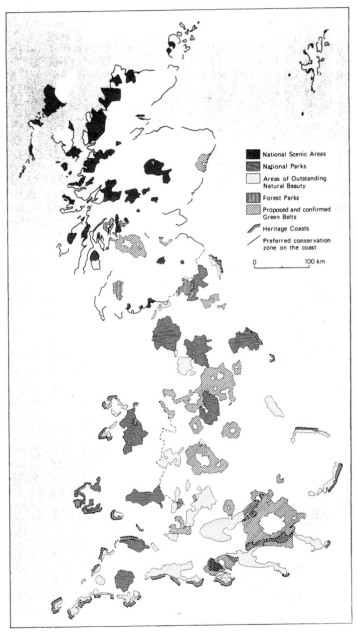

Figure 4.2: Land Designated for Protection in Britain by
Central Government for Reasons of Landscape Value in 1981

to any proposed change of either use or appearance; and, since 1974, this has included control over demolition. Cullingworth reports that, within a few years of the passage of the Act, over three thousand conservation areas had been established (1979, p. 163), ranging from small villages to the whole of the centres of such historic towns as Chester and St. Andrews.

In addition to the powers exercised by local authorities, central government also influences the pattern of land use in several important respects, both directly and indirectly. One of the most important direct influences has been the National Parks and Access to the Countryside Act of 1949. Under this Act a variety of different types of land have been formally recognised as being deserving of a special degree of land-use control. Ten National Parks have been designated in England and Wales, and five National Park Direction Areas were identified in Scotland. The detailed control of land use within the English and Welsh parks is the responsibility of National Park Authorities, which are composed of the nominees of central government and members from the county councils in whose areas the park lies. They are obliged to draw up plans for the park, which must be approved by central government; and they have a special duty to preserve amenity and promote the enjoyment of the public within them. Central government also requires the statutory undertakers to plan their activities in the Parks with regard to amenity. In Scotland there were no Park Authorities, but all applications for planning permission within the Direction Areas had to be notified to central government. However, the Areas were replaced in 1981 by a larger number of National Scenic Areas (Figure 4.2); and only proposals which are opposed by the Countryside Commission for Scotland, but which the local authority wishes to grant, must now be notified. Other scenic areas in England and Wales have been recognised by central government as Areas of Outstanding Natural Beauty, and are subject to similar protection to that of the National Parks, albeit in the absence of special Authorities; and those parts of the coast which do not fall into either of the two previous categories, but which in the view of the Countryside Commission (for England and Wales) and the local authorities justify extra protection from development on scenic grounds, have been designated as Heritage Coasts (Figure 4.2). There are also a large number of smaller areas

throughout Britain which have been recognised as
National Nature Reserves or Sites of Special
Scientific Interest on the advice of the Nature
Conservancy Council, where any planning authority
which intends to permit development against the wish
of the Council must advise central government of the
fact. Altogether, almost a fifth of the country is
covered by these various types of special
protection, two-thirds of which is in the uplands.
     At first sight it may not appear that the extra
controls in these areas have had a very significant
effect. Government, the statutory undertakers,
private firms and the Forestry Commission have been
allowed to carry out major defence, industrial and
forestry projects, and to build reservoirs, while
farmers have altered the landscape substantially in
several designated areas by ploughing and fencing
moorland. Furthermore, many Sites of Special
Scientific Interest have been damaged by farmers and
others, largely through ignorance of their status.
The ability of the Countryside Commission and the
Nature Conservancy Council to prevent land-use
damage in these areas may have been strengthened by
the <u>Wildlife and Countryside Act 1981</u>; but the <u>Act</u>
enables farmers to claim compensation if grants for
the improvement of such land from the Ministry of
Agriculture are withheld at the request of either of
the two bodies. On the other hand, a variety of
studies have shown that permission for the
construction of housing is generally, though not
always, less easy to obtain in the protected areas
than in others (Blacksell and Gilg 1981, pp.
138-149).
     Other forms of direct central government
control over land use include its power to schedule
'ancient monuments' and list buildings of special
architectural or historic interest. Such action
confers protection upon these objects, which range
from individual houses to medieval monasteries and
Hadrian's Wall, from alteration or demolition.
Moreover, through the Forestry Commission it has
licensed the felling of commercial timber since
1951, and thus can insist on the deferred or phased
clearance of privately-owned woodland, and its
subsequent replanting. It was also empowered by the
<u>New Town Act 1946</u> to designate new towns and to set
population targets for them; and it chose sites for
twenty-nine between 1946 and 1973, before cancelling
the last in 1976.
     The indirect influences of central government
are also important. Amongst these the level of
support to local authorities for such activities as

house building and derelict-land clearance, and the requirement that councils seek central-government approval before embarking on capital spending have meant that the scale, if not the detailed location, of local-government activities which have involved changes in land use has been closely controlled by the centre, with consequent effects upon the rate of those changes. Similarly, the level of financial support which has been offered to agriculture since the Agriculture Act of 1947, and to forestry, both private and that of the Forestry Commission, has influenced the allocation of land among the competing uses in rural areas to a considerable extent. It is likely that the areas under all rural uses, but especially upland and hill farming, dairying and forestry, would have been much smaller than they were in the early 1980s in the absence of these supports. Moreover, farmers would not have drained so much marsh or improved so much moorland if they had not received grants from the government for doing so. Indeed, it can be argued that the effect of the subsidies, and of the import restrictions and tariffs, on a variety of farm products has been to cause an increase in the intensity of rural land use in general. Thirdly, central government funds a variety of regional organisations which have important powers to acquire, develop and reclaim land. One of the first of these was the Highlands and Islands Development Board in Scotland, which was established in 1965, but which, in spite of its powers of compulsory purchase over underused rural land, has had little effect upon the competition for land within its area. More important have been the Scottish Development Agency and its Welsh counterpart, which were set up in the mid-1970s and have taken over the responsibility for the reclamation of derelict land from the local councils. The Scottish Agency has also been responsible for much of the redevelopment of the eastern area of Glasgow following the clearance of large tracts of unfit housing by the city council, and the closure and abandonment of much industrial and transport property; and, more recently, the London Docklands have been placed in the hands of a similar body (McAuslan 1981, pp. 248-255).

    All this may suggest that government - and especially central government - has taken increasing control over land use in Britain during the present century; and that would be true. However, the statutes contain but an outline of the controls which can be applied; and much of their

effectiveness is dependent upon the nature of the
advice which is issued to planning authorities,
government agencies and the statutory undertakers in
the form of circulars and advice notes, and the way
in which the call-in and other decision-making
powers of central government are used. It is,
therefore, of relevance to note that, since the
election of a Conservative government in 1979 and
the onset of a severe economic depression, there may
have been some easing of controls. For example,
suggestions that the Green Belts may be in need of
radical review have been canvassed; greater emphasis
has been given to the need for housing, if necessary
on greenfield sites in rural areas (Herington 1982);
and the 1960 policy of opposing such development in
the open countryside has been modified.
Notification requirements of aggregate working and
coastal development have also been reduced (Scottish
Development Department 1981); and there has been
some reluctance to designate further Areas of
Outstanding Natural Beauty, where these might
conflict with mineral extraction (Clark 1982).
Furthermore, the range of development which is
allowed without planning permission under the
General Development Order has been widened in areas
outwith the National parks, Areas of Outstanding
Natural Beauty and conservation zones. To some
extent these shifts in policy represent a devolution
of power to the local planning authorities, but they
may also amount to a reduction in the degree of
control over the allocation of land between
competing uses.

THE CASE OF STRATHCLYDE

     Some idea of the present system of land-use
control may be obtained from an examination of the
situation in Strathclyde Region. The area is
centred on the Clyde valley and Glasgow, and it
houses about half of the population of Scotland; but
it also includes remote and scantily-inhabited
sections of the Highlands and the Southern Uplands.
Since the Second World War the decline of the
traditional industries of coal mining, iron and
steel, heavy engineering and shipbuilding, and,
during the late 1970s and early 1980s, of some of
the newer industries which had been persuaded to
establish in the Region, and especially car
manufacturing, has been accompanied by a decline in
the population from its peak of 2,580,000 in 1961 to
2,400,000 in 1981. But more spectacular has been

the demolition of large areas of congested and unfit
tenement housing around the centre of Glasgow,   and
particularly in what has become known as the Glasgow
Eastern Area.   Large tracts, covering ten percent of
the city, have been left empty as the population has
moved   away   to huge council estates,   housing up to
forty thousand people each, on the edge of the city,
and to the New Towns of East Kilbride,   Cumbernauld
and   Irvine.   As a result of the   success   of   these
towns in attracting not only residents, but also new
industry,   the population of Glasgow has fallen   so
rapidly   from   its   peak   of   1,140,000   in   1961 to
763,000 in 1981 that the government was persuaded in
1976   to   cancel   a   fourth   new   town   which it had
earlier   designated in the Region - at   Stonehouse -
and, through the Scottish Development Agency, to set
up the Glasgow Eastern Area Renewal scheme.   Much of
the   responsibility   for   the   regeneration   of   the
worst-affected   part of the city has passed from the
local   authority,   for whom the task seemed to have
become   too great,   to this body.   At the same   time
concern   was   being   expressed by central government
that suburban and New Town expansion had transferred
large quantities of prime agricultural land to other
uses.   Between 1962 and   1976   10,350   hectares   of
farmland   was taken over for urban development,   and
between   1971 and 1976 sixty   percent   of   what   was
transferred   was   prime   land.   Furthermore,   about
three   times as much land was taken from farming for
afforestation   (Strathclyde   Regional   Council 1979a,
pp.   28-33);   and all this represented a total loss
of   about   five   percent   of   the   Region's farmland
during the 1960s and 1970s.
      Land-use   change   in the Region since   the   late
1970s   has   been governed by a variety of influences
and   controls.   In   addition   to   the   general
legislation   described already in this chapter,   all
Scottish   planning authorities must have   regard   to
the   National   Planning   Guidelines   (Scottish
Development Department 1977 and 1981), which contain
instructions   from   central   government.   These are
sent   out by the Scottish Office,   and are   somewhat
different from those which are issued to authorities
elsewhere   in Britain.   They comprise general advice
about   certain   types   of   land,   and specific advice
about particular sites.   Thus,   at the time when the
first   Structure   Plans   were   being   drawn   up,
authorities   were advised that   there   should   be   a
presumption   against   taking prime agricultural land
for development or quarrying unless it was to be for
industry   in one of the preferred zones on the coast
(that is,   not in the Preferred Conservation Zone in

Figure   4.2),   or for industrial developments which
were of national importance,   such as oil and power,
or unless all sites on poorer land had proved to be
unsuitable.   They were also requested to identify a
few suitable sites for large-scale industry, each of
which was to cover a hundred hectares or more.   In
the case of Strathclyde it was indicated that these
should   be   to   the   northeast   of   the   Glasgow
conurbation,   in the area of the Strathclyde-Lothian
regional boundary, and in north Ayrshire, preferably
on or near the coast.   The _Guidelines_ reminded
authorities   that they should coordinate their plans
for land use with those of the Forestry Commission,
the Countryside Commission for Scotland (in the case
of the National Park Direction Areas and Areas of
Great Landscape Value) and the Nature Conservancy
(in the cases of the National Nature Reserves and
the Sites of Special Scientific Interest).   More
than half of the Strathclyde Region fell within one
or more of these designations in 1979 (Strathclyde
Regional Council 1979a, p. 55).   In addition, broad
areas of the country were indicated in which the
Scottish Office would be 'favourably', 'normally' or
'only exceptionally' in support of sand and gravel
extraction.   A revised set of _Guidelines_ was issued
in 1981, but the major lines of central government
land-use policy remain largely unchanged.
     It was in this context that the second major
control   over   land   use - Strathclyde Regional
Council's **Structure Plan** - was drawn up, in which
the aims and policies of the Council are set out.
The primary aim is

          "to   assist   employment   by   increasing   the
          attraction of the Region for investment and to
          reduce   the   flow   of employment and population
          from   the conurbation by completing the process
          of urban renewal and regeneration" (Strathclyde
          Regional Council 1979b, p. 7).

More particularly, there is to be a preference for
residential development on infill and redevelopment
sites within urban areas, rather than on peripheral,
greenfield land; and it is recommended that the new
towns   should   only   build   for   their   existing
populations or in connection with any growth in
employment within or around them. Maximum numbers
of new houses are set for each part of Glasgow and
for each of the Clyde-valley towns for the period to
1983, and some additions to and deductions from the
supply of land already allocated for housing are
specified for a number of settlements (pp. 25-29).

There is also a list of the settlements in which additions to the supply of land for industry would be supported (pp. 39-44), but no extension of sand and gravel extraction or quarrying is to be allowed (p. 62). Major office building is to be restricted to the central area of Glasgow (p. 68). Lastly, the Structure Plan discourages development outwith existing communities in areas of "regional scenic significance", and the spread of the built-up area into the Green Belt (pp. 56-58). Thus the emphasis has been placed upon the concentration of development in the existing towns of the Clyde valley, and especially in Glasgow, where the rebuilding of the urban fabric is to enjoy a very high priority.

The third major control over land-use change is the Local Plan. These will eventually be drawn up for all parts of each District, but in Glasgow attention was paid at the start to areas with severe problems. One of these is Dalmarnock, which lies immediately to the east of the city centre and north of the Clyde. It contains a mixture of land uses. Two-fifths of the housing in 1978 was in tenements which had been built before the First World War, a quarter was in inter-war blocks of council flats, and much of the rest was in multi-storey blocks of flats built by the city in the 1960s. In addition, there were factories, a sewage works, a disused electricity-generating station, and several areas where buildings had been demolished, but not replaced. Some of the land had been 'blighted' for many years by a proposal to build a motorway across it. Population density in the residential areas was high at 430 people per hectare, but it had been higher in 1971 before some of the older housing had been demolished, when the population had been about 9,000. By 1977 it had fallen to 5,500, and was expected to decline still further. There were also 4,200 jobs in the area in 1977, four-fifths of which were in manufacturing (Glasgow District Council 1978, pp. 3-5, 13-17). Altogether, twenty-three percent of the small total area of 101 hectares was in industrial use, nineteen percent was idle or under vacant premises, and thirteen was covered by housing (Figure 4.3).

The Local Plan for Dalmarnock (Figure 4.4) is still dominated by the motorway link, although construction of it was not expected to begin before 1983, if at all. The link bisects the area into a southern section, in which the remaining substandard housing and the electricity works are to be demolished, and in which the vacated land is to be

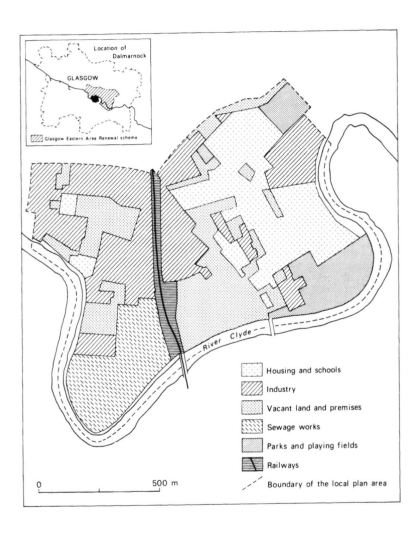

Figure 4.3:   Dalmarnock Local Plan Area - Land Use in 1981

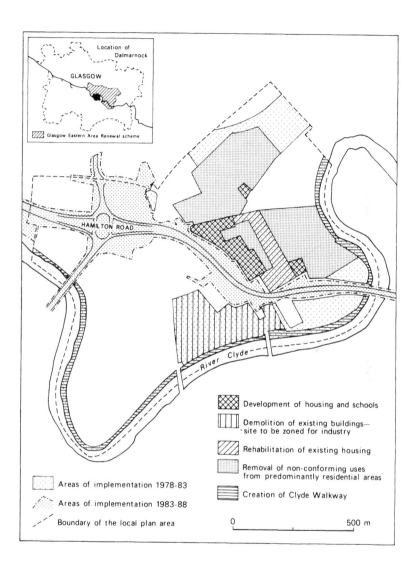

Figure 4.4: Dalmarnock Local Plan Area - Proposed Future Pattern of Land Use

zoned for industry, and an area of mixed land use to
the north. Some new housing was to be added on
empty sites in this northern section by 1983, and
other housing improved; but the exclusion of the
remaining nonconforming uses will not be tackled
until later. If the whole plan is carried out all
the vacant sites in the area will be developed or
used for public open space, the housing will have
been improved, and the general appearance will have
been softened by the planting of trees. However, it
is not expected that the population will ever again
reach the level of 1971, or that in general the land
will be used quite as intensively as it was during
the first sixty years of this century. Whether or
not the plan is achieved will depend in part upon
decisions by central government about the motorway
building programme, and in part upon those of the
Scottish Development Agency and Glasgow District
Council about the house-building and
environmental-improvement programmes. But they will
also depend in large measure upon the willingness of
private industry to take up the sites which will
become available, albeit with the assistance of some
considerable financial incentives which are on offer
to businesses on those sites from central
government.

CONCLUSION

        All this suggests that the British government
has both a clear policy with regard to land use, and
a well-structured, hierarchical machine for carrying
it into effect. Major industrial developments of
national significance are to be given priority over
almost all other land uses, but otherwise prime
agricultural land, areas of recognised scenic and
historic value, and sites of natural and scientific
interest are to be protected. The continued
expansion of forests is not to be held back
unreasonably on land which is of only limited value
for other purposes; but the growth of settlements is
to be accommodated by the use of land within the
built-up area, especially if it is vacant, rather
than relying solely on peripheral extension. All
this is to be carried out through consultation
between the planning authorities and such bodies as
the Forestry Commission, the Nature Conservancy
Council and the Countryside Commissions, and by the
system of development control based upon local
plans. District Councils are to determine the

detailed   arrangement   of   land   uses   within   the
aggregates   laid down by the Structure Plans;   and a
variety   of   requirements   for   the   notification of
proposals   to,   and the call-in powers possessed by,
central   government   seek   to ensure that matters of
national   importance remain within the control of the
centre.   Furthermore,   in so far as this system of
regulation   places   the control of land   in   the
hands of elected bodies, it would appear to give the
general   public   the   power   to   protect   what   it
perceives   to be of greatest value in both the rural
and the urban scene.

Notwithstanding   all this,   land-use change   in
Britain   has   been ridden by controversy both before
and   since   the   major   controls   became   firmly
established   in   the   late 1940s - controversy which
has   stemmed from the lack of a single view   amongst
the   interested   parties   as to what types of change
should   be allowed,   and the lack   of   precision   in
government policy with regard to land-use priorities
which   has arisen in consequence.   Should government
be   continuing   to   emphasize   the protection of the
country's   agricultural potential when it has become
largely   self-sufficient in temperate foodstuffs and
is   part   of   a   trading   bloc   which   suffers   from
highly-priced   surpluses   of most of Britain's major
farm products? Should the Ministry of Agriculture be
offering   grants to farmers to improve their land at
the   same time as bodies such as the   National   Park
Authorities   are   obliged to compensate them for not
proceeding   with the ploughing of moorland?   Is   the
price   of   land for forestry being kept unreasonably
high as a result of government subsidies to hill and
upland farmers; and should public access to National
Parks   and National Scenic Areas   be   restricted   in
order   to preserve their character at the expense of
recreational   opportunities for the general   public?
These,   and many other conflicts,   remain unresolved
and   indicate that there is no broad consensus as to
what   the detailed priorities in land use should be.
The   Select   Committee   on   Land   Use   Resource   in
Scotland's   call   (House   of   Commons   1972)   for   a
national   land-use   plan   was   rejected   by   the
government of the day.

Thus, we may now return to Kirk's claim that in
Britain "land tends to be put to its most profitable
use".   Much   evidence can be brought to support it.
Firstly,   while nine-tenths of the country, and much
of its economy, remains in private hands, control of
land use is chiefly a negative operation of refusing
permission   for change.   But in a period of economic
recession,   such controls are   more   likely   to   be

replaced in effect by a scramble among local authorities for any economic development which is available. Moreover, because of the dominance of private landownership, government, both central and local, is obliged to pay for land at prices which are set by the market, and therefore is strictly limited in the amount over which it can afford to exercise positive control. Secondly, much land-use change occurs in the pursuit of profit despite the existence of controls or, at least, strong public opposition to it. Farmers retain great freedom to alter the quantity and arrangement of improved land, marsh and moor; mining companies have managed to win permission for the extraction of, for instance, limestone and potash from National Parks; industry has obtained the right to refine petroleum and build oil rigs in its preferred locations, in spite of their designated scenic status; and developers have carried the tide of bricks and mortar, apparently relentlessly, into the countryside, in spite of the legislation of the 1940s (Coleman 1976). Government has shown its willingness to override Development, Structure and Local Plans in favour of developers, owners and producers, and has thus severely curtailed the credibility of the system of control. Conversely, it has required the payment of substantial subsidies by the state to keep much of the Highland zone in agricultural use, and more subsidies to transfer large areas within it to forestry; while private developers have only been tempted back onto many of the expensive, inner-city sites by the provision of advance factories, investment grants and other inducements by government. Nevertheless, considerable changes have occurred in the broad aggregates of land use as a result of these interventions; and, where planning permission has been sought by private firms and individuals, there has been a much lower chance of success in areas of designated scenic value than in other parts of the countryside. We may conclude that much more land, and much more of the most valuable areas for agriculture, nature conservation and scenery, would have been drained, afforested or built upon if the 1947 Town and Country Planning Act and other legislation had not been passed. Thus, Kirk's claim cannot be completely sustained for the period between 1947 and 1979, although there are indications that it may be more correct in the immediate future as the Conservative government eases some controls. However, until it is clear that a considerable easing has occurred - and there has been no significant reduction in the statutory

powers of government - it must be concluded that the
land problem as perceived by the British - and they
have perceived it in many different forms and from a
variety of viewpoints - has been tackled with the
aid of considerable powers and resources for at
least thirty-five years, to marked effect. That it
has not always been tackled to the satisfaction of
the entire community is a measure of the conflicts
embodied in the simultaneous pursuit of many goals
with respect to the use of land, few of which are
independent of each other, and of the variety of
points from which the problem has been viewed.

Chapter 5

LAND-USE REGULATION IN JAPAN

"in... such an economically buoyant society,
considerations of land use planning are likely
to take a bad second place in competition with
the overwhelming desire to exploit every
economic opportunity; put at its bluntest, most
Japanese decision-makers - and perhaps most
Japanese people - have little time for planning
at present." (P. Hall [1977], The World Cities
(2nd ed.), Wiedenfeld and Nicolson, London, p.
238).

INTRODUCTION

Of all the developed economies included in this
study, Japan is the one which appears at first sight
to have the most serious land problem. Not only is
the density of population there far in excess of
that in the United Kingdom, let alone of those in
Eastern Europe or the United States, but also the
number of people in the country has been growing
faster during recent decades, and the area of
farmland declining more rapidly (Table 1.3).
However, the economy has expanded to such an extent
that Japan is now one of the richest societies in
the world, and is very highly developed according to
all the indices used in Table 1.1. In addition, the
diet, which in the past was low in protein, lacked
variety and was often inadequate, has improved
greatly; and production of the staple crop - rice -
has been in excess of demand since 1969. All this
has been achieved in the absence, until the late
1960s, of any effective, nation-wide system of
control by government over land-use change, such as
that which has existed in Britain during the period
since the Second World War - an absence which must

call in question the need for similar controls in
countries with less acute land problems. But,
before any international comparisons can be drawn,
we shall proceed as we did in the case of Britain
with an examination in detail of the land resources
of Japan, of the attitudes of the Japanese in
general, and of some of the pressure groups in
particular, to land, and of the evolving role and
policies of the Japanese government with regard to
land use in the country, both as a whole and in one
of its most populous parts - southern Kanto - since
the modernisation of the country began after the
Meiji restoration of 1868.

THE SUPPLY OF, AND DEMAND FOR, LAND

        Although Japan is a larger country than
Britain, the area suitable for agriculture and
settlement is much less. Sixty-one percent of the
land is either mountains or volcanoes (Prime
Minister's Office, 1981 pp. 2-3), and sixty-eight
percent is covered by thin, stony and immature
soils. Only a quarter of the country is in slopes
of less than fifteen degrees, and some of these
areas, including much of the Kanto Plain and the
lowlands of south-east Hokkaido, are covered with
light, porous soils derived from acidic volcanic
ash, which are easily leached, lack organic matter
and are generally infertile. Only twelve percent of
Japan is covered by alluvial soils of moderate base
status, silt or clay content, and organic matter
(Dempster 1969, p. 57-61). It is true that many
steep slopes have been terraced for paddy
cultivation in the past, especially in the south of
the country; but it is difficult, if not impossible,
to create fields of suitable size for mechanised
cultivation ·on such sites, and forest clearance in
the absence of terracing brings risks of soil
erosion, river sedimentation and downstream
flooding.
        Furthermore, the demands upon the limited areas
of usable land have been growing. Population has
increased almost four-fold since the middle of the
last century, and manufacturing and foreign trade
have grown to a much greater extent. Much of this
took place before the Second World War, but the
contribution of the last thirty years has also been
striking. During the latter period huge factories,
refineries and port installations have been built;
and increasing wealth has permitted both a doubling
of the area of housing available per capita and the

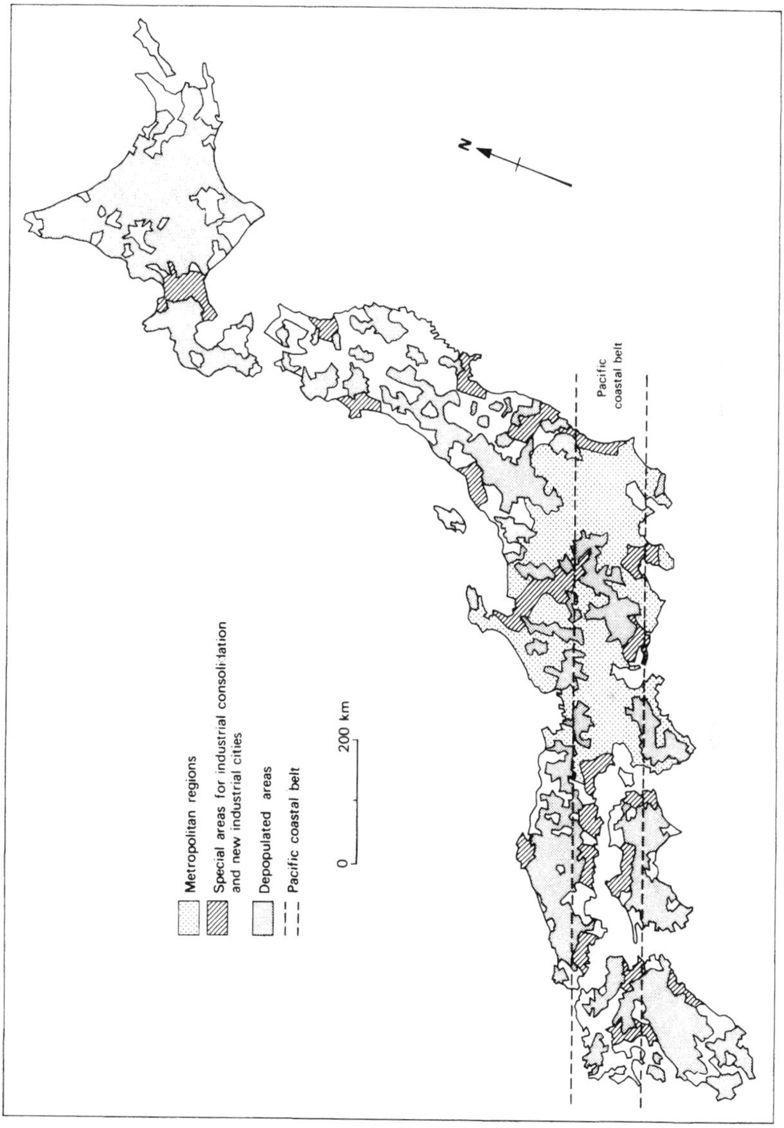

Figure 5.1:

Some Problem

Areas in Japan

substitution of meat and dairy products for some of
the carbohydrates in the Japanese diet - all changes
which have made extra demands upon land.
     However, the incidence of those demands has not
been uniform across the country. After the outbreak
of the Korean War industrial growth was heavily
concentrated in the largest cities along the Pacific
Coastal Belt (Figure 5.1); and suburban expansion
there was driven on by the dual movement of people
into   the   Belt   from   other,   rural   parts   of the
country,   and out of   the   congested   inner   cities
within it.   Much of the growth of settlement was
inevitably on the gently-sloping and flat lands of
the Kanto Plain and the other lowlands, which had
traditionally been major areas of agriculture; but
the supply of suitable land for mills and refineries
was   so limited that 60,000 hectares of the sea were
reclaimed between 1965 and 1975,   and landfill
operations advanced from areas under ten metres of
water to some with depths of up to fifty metres
(Sakiyama   1979,   p.   24).   At the same   time
structural change in the economy was of such a pace
that the emigration from rural areas caused absolute
falls in population there, notwithstanding their
high rates of natural increase. Between 1960 and
the mid-1970s forty-five percent of the country
experienced population decrease at the rate of at
least ten percent within a five-year period, and was
designated   by the government as   "depopulated"
(Figure 5.1).   Much agricultural land was abandoned
in these regions. More recently, the desire to
migrate to the Pacific Belt has waned, for, although
employment has continued to expand on the lowlands
between Kanto and northern Kyushu, growth has also
been experienced in other parts of the country
(Itakura 1980,   pp. 83-86), and income levels in
different areas have converged (Mera 1977,   pp.
460-462). Nevertheless, the increase in population
has continued in the three metropolitan regions
(Figure 5.1), and has been greater than in the rest
of the country throughout the 1970s, while much of
the new industry in other areas has been located in
the larger towns with the result that emigration and
farmland abandonment have continued in the smaller,
ageing communities in mountainous and remote regions
(Kakiuchi and Hasegawa 1979, pp. 49-50).
     When faced with problems of land shortage in
the past the Japanese have usually responded in a
variety of ways. Poorer or more steeply-sloping
areas have been reclaimed, and the limit of
cultivation has been pushed off the lowlands and

into the upland fringes or onto land reclaimed from shallow coastal waters. Paddy irrigation has been extended in order to increase yields of rice; and double-cropping has been practised, in which rice has been followed later in the year by barley, rice, vegetables or wheat, where the climate would permit it. Holdings have also been subdivided between heirs. Some of these responses to the growing pressure on land have been manifest again since the Second World War. For example, notwithstanding the transfer of much land from agriculture to urban uses during the period, the cultivated area increased until it reached a peak of 6,086,000 hectares in 1961. Similarly, the area under paddy grew until it reached 3,441,000 hectares in 1969. At the same time the price of land soared at a far faster rate than that of the general level of prices - though the rate was similar to that of the Gross National Product until 1975 - so that by 1980 the price for paddy was more than five times what it had been twenty years earlier (Prime Minister's Office 1981, p. 120).

Some indication of the present intensity of land use in Japan is given by comparisons with other developed countries. For example, during the late 1970s the application of artificial fertiliser per hectare on Japanese farms was the highest in the world, being a quarter more than that of the next country, the Netherlands, more than three times that of the United Kingdom, and nine times that of the United States (Glowny Urzad Statystyczny 1980a, p. 179). Secondly, the average price of dry-field land was about three times that for similar land in the United Kingdom, and twelve times that for the United States, and for paddy it was four and twenty-five times as expensive as land which could be considered to be comparable (Ogura 1979, p. 781). More generally, Mills and Song (1977, p. 8) have calculated that, whereas the value of all land in the United States in 1975 was equal to about seventy percent of the Gross National Product, that in Japan was worth 330 percent. As a result the average Japanese worker in 1970 was obliged to work for more than six years before he could buy 150 square metres of land within forty minutes commuting time of central Tokyo, while in France he would have been obliged to work for less than three years, and in the U.S.A. only forty-five days for the same land in a comparable location (Tokyo Metropolitan Government 1976, p. 2); and it should be remembered that the greatest inflation of land prices in Japan occurred after that year.

Government forecasts suggest that these pressures will continue to increase. The population is expected to rise by more than twenty million between 1980 and 2010, before it stabilises; and if the cultivated area remains at the same level, the utilisation ratio will have to rise from 103 to 112 percent if the present level of self-sufficiency in farm products of seventy-three percent, which has already declined sharply since the 1950s, is to be maintained (Ministry of Agriculture 1980, p. 3). However, any continuation of the long-established trend towards the westernisation of diet could make even this degree of self-sufficiency difficult to achieve; and, although the Agricultural Policy Council suggested in 1980 that the calorific value and balance of the Japanese diet were almost ideal, the government's ability to halt the increasing demand for animal products (which require much greater areas of land for the production of fodder crops than for crops of equivalent food value which are grown for direct human consumption), even at the very high prices which it has fixed for them, is very limited (Oriental Economist 1981, p. 39). In addition, there is little scope for the conversion of forest land to other uses. Not only is most of it on slopes which are too steep for many other activities, but also the degree of self-sufficiency in forest products - at forty percent - has shown a dramatic decline since the 1950s. Lastly, satisfaction of the continuing demands for better housing, more jobs, increased water supplies and improved transport networks will also require land (National Land Agency 1979, pp. 66-72).

ATTITUDES TO LAND

It might be supposed that, in view of these pressures, the Japanese would, like the British, have developed a conservative attitude towards land. But, as Hall (1977, p. 238) suggests, that has not been the case. Although Japanese religion has ancient roots in the worship of rocks and mountains, and although there is a widespread appreciation of the beauty of the wooded mountain scenery of the country (Kaneyasu 1981, pp. 45-46) there does not exist a well-developed romantic view of these areas or of the rustic way of life. Land hunger, the shortages of food, the high rents extorted by landlords, and the peasants' burden of debt - all of which were very present problems right up to the land reform of the late 1940s - have not provided fertile ground for the development of the Britisher's rosy and nostalgic view of rural landscapes amongst either the many millions of Japanese who migrated to the cities after 1950 or those who have remained on the land. Of more importance for most of the people since 1945 has been the need to save sufficient from their incomes to provide, firstly, in a country with some of the world's highest house prices, for the eventual purchase of a modest dwelling; secondly, in the absence of a comprehensive state welfare system, for ill health and retirement; and thirdly, in the absence of any general system of student grants, for their children to attend university. As a result, the rate of saving by the Japanese has been exceptionally high, and this in turn has permitted very rapid rates of investment and economic growth. What is more, the population at large has appreciated that it would only be through such growth that rapidly-rising levels of real income and the achievement of their dreams could become possible; and so little support was shown until the late 1960s for any control over commerce or industry which would have threatened the ability of those activities to expand. Even then, the public was much more concerned with the immediate impact of growth upon itself in the form of atmospheric and water pollution, and of the diseases which they caused, than upon any long-term possibility of a threat to the agricultural output of the country or to the intrusion of development into rural landscapes. Nor has there been any broadly-based public movement to protect the country's cultural heritage. Approval of the 'new', and preference for it over the 'old', extends from matters of technology to building and land use, and provides an environment which is more favourable to those who wish to develop than to those wishing to conserve. It is of some interest to note that Japan has no institution comparable to Britain's National Trust. The equivalent body in Japan, the Protection Foundation for Tourist Resources, was only established in 1968, and ten years later had a membership of only about two thousand (Tajima 1978, p. 34), while the number of other amenity and conservation groups is small.

The attitude of the farming community towards land is of particular interest. Ogura (1979, pp. 628-637) suggests that, after escaping from the hardships of tenancy which existed before the land reform, farmers have adopted an entrepreneurial attitude towards their holdings. As evidence of

this he cites their reluctance to sell land in spite
of the fact they they cannot, or do not wish to, use
it to the full, and the fact that they have
increasingly abandoned the practice of double
cropping since 1955. All this he deplores as a
breach of faith with the aim of the reform, which
was intended to give land to the tillers, who, in
return, were expected to use it responsibly. The
implication of the criticism is that farmers are
holding on to land as a source of capital growth, as
a means of obtaining inflated incomes from the high
and guaranteed rice prices paid by the government,
and to ensure a cheap supply of food after
retirement, at the expense of an urban population
which, in consequence, continues to live in small
houses and congested environments. Ogura (1979, p.
607) notes, however, that opposition to any
suggestion that farmers should be obliged to use
their land more 'responsibly' is so strong that
there is a consensus among the political parties
against any extension of controls over the intensity
of land use or tampering with the freehold rights of
the family farm.
      And so, despite the Japanese tradition of
strong, central authority, the role of government
has been to encourage, rather than direct, the
changes which have affected land use. Thus in the
Meiji era government promoted improvements in
agriculture, assisted with the establishment of new
industries - largely through the provision of the
necessary infrastructure - and organised land
reclamation projects and the settlement of
Hokkaido. But it left the necessary entrepreneurial
action to private individuals and firms. Similarly,
since the Second World War it has given very high
priority to the growth of industry, and has
encouraged the transfer of factors of production
from less to more-productive uses (Allen 1974, pp.
31-62), but it has made only weak attempts to
determine the distribution of industry and
settlement across the country. Efforts were made to
entice industrial growth away from the three major
conurbations around Tokyo, Osaka and Nagoya in the
1960s to other parts of the Pacific Coastal Belt -
the Special Areas for Industrial Consolidation - the
New Industrial Cities (Figure 5.1), the declining
coalfields, and, in the 1970s, to other parts of the
country; and a large number of planning documents,
such as the various Comprehensive National Land
Development Plans and the National Capital Region,
Kinki and Chubu Development Plans, have spelt out
government hopes for some dispersal. However,

action to achieve this has been limited very largely
to the improvement of infrastructure in the more
slowly-growing areas and the offer of subsidies to
firms moving there. In the same manner farmers were
encouraged in the 1960s to substitute other crops
for rice; and, since a surplus of rice production
developed in 1969, the government has paid them to
reduce the rice acreage, and, more recently, let the
real value of the guaranteed price for rice fall;
but it has taken no more direct action. In fact,
direct action to control the spatial pattern of
development was taken by government, at least until
the 1960s, in only a few, limited cases. For
instance, after the Kanto earthquake and fire of
1923, in which 143,000 people were killed, controls
were imposed on the height and density of buildings
in Tokyo. Also there was legislation in 1936 under
which most of Japan's National Parks were
designated, and several laws were passed to protect
buildings of historic interest. But, until the
prohibitions on the establishment of major new
industries or universities in Tokyo in 1959, few
direct land-use controls existed. However, if
Glickman (1979, p. 257) is correct in his
suggestion that during the 1950s and 1960s Japan was
undergoing a stage of urban growth which other
developed nations, including Britain, had passed
through earlier in the century, it may not be
surprising that the role of government in land-use
planning had not evolved to the same extent,
notwithstanding the very considerable demands upon
the land during those decades.

However, a substantial shift of both public and
governmental opinion did occur in the late 1960s and
the early 1970s, for the Japanese were caught up in
the world-wide concern about the environment and the
finite nature of physical resources. Court cases
against companies which had polluted the environment
and caused disease in several parts of the country
provided a powerful focus for this concern; and
there were complaints about the congestion of the
cities, the increasing dependence of the country
upon imports of food and fuel at a time of world
cereal shortages and rapidly-rising petroleum
prices, and the very substantial increases in the
price of land. Thus many factors contributed
towards a widespread demand that government should
take significant and wide-ranging powers to control
changes in the use of agricultural land and the
pattern of urban development, and to coordinate the
plans for these land uses with those for forests,
parks and nature conservation.

SOME IMPORTANT ACTORS

Nevertheless the political environment surrounding questions of land-use change has been little altered. Notwithstanding the increasing public concern, amenity and conservation groups have remained extraordinarily weak, being small in both numbers and membership, and local and fragmented in organisation (Hase 1981, pp. 18-20, 34-36), community groups have wielded little power, local government has remained largely the agent of the central authorities, upon whom it is highly dependent for its income (McNelly 1972, pp. 194-211), and it has been the backers of the Liberal-Democratic Party - the farmers and big business - who have continued to be the effective controllers of Japanese land policy throughout the post-war period.

The great influence of the farmers is related to their recent history. Following the land reform of the late 1940s most of them were relieved of the severe burdens of tenancy, and became owner-occupiers. As such they have proved to be conservative in outlook, and have given strong support to the Liberal Party, which has provided the Japanese government, either on its own or in coalition with the Democrats, since 1949. Agricultural cooperatives, which exist in every village, have been a major channel for the collection of financial contributions for the party, and many cooperative officials have been elected to the Diet as government supporters. Moreover, immediately after the Second World War more than half of the population lived in rural areas, and no redistribution of constituencies for the Diet occurred between 1947 and 1983 to take account of the huge shifts of population over that period. Thus farming areas are grossly over-represented, and the 1983 alterations, which introduced proportional representation into the elections for the upper house of the Diet, have done little to alter the situation. It is therefore not surprising that

"the Japanese government has probably never made any major changes in agricultural policy without first consulting with national cooperative leaders"

or that, during the 1960s, the National Association of Agricultural Cooperatives (Nokyo), to which almost every farmer in the country is affiliated, managed to persuade the government to increase the guaranteed price for rice at a greater rate than was necessary to maintain the parity of farm with other incomes, despite opposition from the Economic Planning Agency and the Ministry of Finance (Donnelly 1977, pp. 152-159). Indeed, so great has the National Association's success been, even since the production of rice has been in surplus, that its price rose to four times that on the world market in the late 1970s (Coyle 1981, p. 3); and, it has been calculated that as a result of this protection not less than two-fifths of the output in 1976 was the product of government support, and that at least an equal quantity of imported rice was excluded (Bale and Lutz 1981, pp. 8-22). Moreover, the Japanese government has been subsidising the export of rice since 1969, to the extent of selling it at less than world prices, and has been paying farmers handsomely to substitute other crops in its place - crops which also could be bought at lower prices on the world market - and it plans to go on doing this until at least the end of the 1980s (Coyle 1981, pp. 5-14). Much of this expenditure is defended by the government on both social grounds - the maintenance of farm incomes - and in the interest of sustaining as high a level of self-sufficiency in food production as possible; but the desire to reduce the subsidies has also been strong since the early 1970s, and their continuingly-high level is an indication of the influence of the farming lobby.

There is also a close relationship between big business and government in Japan. Directors in the large firms frequently move into cabinet posts, and vice versa; and there is regular, formal contact between the Ministry of International Trade and Industry (MITI) and industrialists for the exchange of opinions (Morishima 1982, pp. 188-190). Big business has made very large donations both to the Liberal-Democratic Party, and to the factions within it; there are frequent allegations of bribery involving business interests and ministers; businessmen make up a large majority of the Liberal-Democratic members of the Diet; and it is generally believed that no one can become Prime Minister without the approval of the <u>Zaikai</u>, or large industrial-commercial-banking conglomerates (Yanaga 1968, p. 33). Both government and industry have been eager to achieve rapid economic growth, and it is hardly surprising that they have cooperated closely during most of the post-war period. However, it has been clear that the

emphasis of government policy has been on growth,
rather than on the regional dispersion of
production; and the impact of its regional policies,
which have been supposed to encourage industry to
move from the congested Pacific Belt to more
peripheral locations, has been weak. In this it has
reflected the desire of the major industries to
occupy coastal sites or to be close to the seat of
government. Nevertheless, the government was
obliged to give way to public outcry in the early
1970s and enact effective legislation to control
pollution, against the wishes of both the
organisation to which all major businesses in Japan
belong - **Keidanren** - and of its powerful supporter
inside government - MITI (McKean 1977, pp.
201-238) - and has introduced some significant
controls over the transfer of land to industrial and
other uses, especially since the late 1960s.

CONTROLS OVER RURAL LAND

Controls over the conversion of agricultural
land to other uses have actually been available
since the passage of the **Agricultural Land Act** of
1952. Although the **Act** was designed primarily to
protect the achievements of the land reform, by
imposing on the governors of the regions
(prefectures) the duty to control the transfer and
tenancy of cultivated land, and thus prevent the
reappearance of impossibly-tiny or illegally-large
farms, it also permitted them to control the
transfer of farmland to other activities. However,
decisions were left to the discretion of the
governors and their professional advisers; and many
of them were quite willing to allow such transfers,
especially where this facilitated the establishment
or growth of manufacturing industry in their
prefectures. In the event, such was the pace and
uncoordinated manner of transfer that in 1959 the
Ministry of Agriculture issued **Standards for
Permitting the Conversion of Farm Land** as a guide to
the prefectures, but these also failed to institute
an effective system of control, with the result that
much land was converted to other uses in the 1960s
in a sporadic and untidy manner, especially in the
Pacific Coastal Belt.
Effective protection of agricultural land from
such development was not provided until the passage
of the **Agricultural Promotion Areas Act** of 1969.
This **Act** required the designation of Agricultural
Promotion Zones in all prefectures, and of

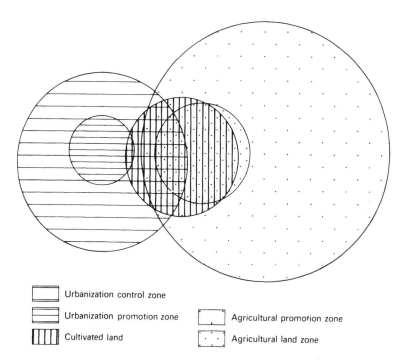

Urbanization control zone

Urbanization promotion zone   Agricultural promotion zone

Cultivated land   Agricultural land zone

Figure 5.2: Overlaps in the Zoning of Cultivated Land in
Japan in 1981. (The sizes of the circles and of the overlaps
between them are proportional to the areas of land involved.)

Agricultural Land Zones in all municipalities within
the Promotion Zones. According to the legislation,
a Promotion Zone must be a rural area, and must
include land which is usable for, or related in its
use to, agriculture, although it need not
necessarily be under cultivation, or indeed ever
likely to be. The designation of such Zones is the
responsibility   of   the   prefectural   governor.
Permission to convert land within a Zone to other
uses is also a matter for the governor, except where
the area is in excess of two hectares, in which case
the decision   is   made   by   the   Minister   of
Agriculture. In either case it is only given in
normal circumstances where the new use would be of
an agricultural or rural nature. The Land Zones, in
contrast, are made up very largely of cultivated
areas. Conversion of land within them to other uses
is not usually permitted; and parcels which have

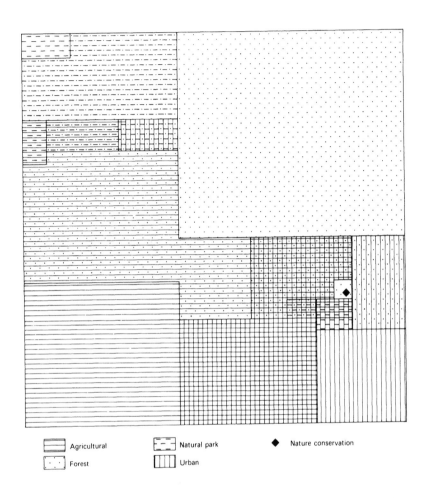

Agricultural   Natural park   ◆ Nature conservation

Forest   Urban

Figure 5.3: Overlaps in the Zoning of Land in Japan in 1981.
(The square represents the entire land area of the country.
Land falling within one zone is shown by a single shading;
land falling in two zones by a double shading, etc., except for
the areas of nature conservation, which are too small to be
shown at the same scale as the other zones.)

fallen out of cultivation may be transferred by the
mayor of the municipality to other farmers. By 1976
17,600,000 hectares, or forty-seven percent of the
country, lay within the Promotion Zones, and
5,600,000 hectares of this, including more than
four-fifths of the cultivated land, lay inside the
Land Zones.

The status of that part of the cultivated area
which lies outwith the Land Zones is more
complicated. Some lies within the Promotion Zones,
some within the parts of those Zones which are also
within what are called Urbanisation Control Zones,
some outside the Promotion, but within the Control
Zones, and about 215,000 hectares, or three percent
of the cultivated area, is included in Urbanisation
Promotion Zones (Figure 5.2). Urbanisation Control
and Promotion Zones were established by the 1968
City Planning Act, and are made up of land around
urban settlements which, respectively, will not be
required for urban settlement within ten years and
on which sporadic building should be prevented, and
that which is already built up or will be needed for
urban use within that period. Agricultural land
within an Urbanisation Promotion Zone may be
converted to other uses without permission, and it
should be noted that the national government has
been quite willing to enlarge these Zones when land
has been required for urban development. However,
ninety-two percent of the agricultural land within
them at present is of the poorest type for
cultivation (National Land Agency 1981, p. 36).

The Agricultural Promotion Zones also overlap
with two other types of zone (Figure 5.3). By far
the larger of these is the Forest Zone, in which lie
lands which are either wooded or which are scheduled
for afforestation, and are of at least thirty acres
in extent. Forest Zones cover two-thirds of the
country, and twenty-three percent of Japan lies
within both the Forest and Agricultural Promotion
Zones. The other overlap is with the Natural Parks,
which cover about fourteen percent of Japan, of
which about a fifth is common to both types of
Zone. In both the cases of overlap any proposed
change of use is subject to the usual controls in
the Agricultural Promotion Zone, and also to the
controls of the other Zone in which it lies.

As with the Agricultural Zones, both the
forests and the parks enjoy varying degrees of
protection, depending upon their importance. Just
as Agricultural Land Zones are recognised within
Promotion Zones, so Reserved Forests and Special
Park Districts have been designated inside the

broader Forest Zones and Natural Parks. Thus, in a
Forest Zone any development which would cover one
hectare or more has been subject to the approval of
the prefectural governor since 1974, and in the
7,080,000 hectares of Reserved Forest development is
usually forbidden for these areas are considered to
be critical to the prevention of erosion, floods and
landslides, the protection of water supplies and the
provision of recreational amenity. Similarly, while
all proposals for development within the Parks must
be approved by the governor, and must be compatible
with the parkland status, development or lumbering
are generally forbidden in the Special Districts.
Control of land-use change should also be
facilitated in these highly protected Zones because,
in the case of the Reserved Forests, four-fifths of
the area is owned by national and prefectural
governments, while almost all the land in the
Special Park Districts is in public ownership; and
the government has been buying up the remaining land
in private ownership in these Districts since 1972
(Environmental Agency 1980, pp. 102-103), albeit
very slowly.
          Thus, there are four major land-use control
Zones in rural Japan; and within each there is a
core area in which change is either subject to
particularly restrictive control or, in the case of
the Urbanisation Promotion Zones, is actively
encouraged. Most overlap to a considerable extent
with one or two others; and all were established
under legislation passed between 1968 and 1974,
although the Natural Parks and the Reserved Forests
had been foreshadowed by earlier laws.

CONTROLS OVER URBAN LAND

          The second major body of legislation with
regard to land use is that dealing with urban
areas. Amongst the earliest city planning measures
after the Meiji restoration were laws in the 1870s
to protect the historic fabric of Kyoto - the former
capital of Japan - and of Tokyo, and to conserve
green areas within the cities. A city planning
commission was established for Tokyo in 1888, and a
Law for the Protection of Old Shrines and Temples
was passed in 1897. The National City Planning Act
of 1919, which consolidated the earlier legislation,
allowed for the designation of Preservation
Districts with a view to conserving historic
landscapes; and, after 1929, areas of scenic beauty
elsewhere in the country were also identified for

protection. However, the areas affected were small,
and the protection afforded was not very effective
(Tajima 1978, p. 33). The 1919 legislation also
gave government power to readjust land-use and
ownership patterns over limited areas of cities in
order to improve traffic flows, or to provide other
amenities such as open space. Between 1920 and 1945
six hundred square kilometres of confused and
congested urban settlements were readjusted, largely
after the Kanto earthquake; and a much larger area
has been redeveloped since the Second World War,
including land along the _shinkhansen_ railway in
Tokyo, Yokohama and Osaka. More recently there has
been further legislation to protect historic
landscapes within towns. The _Old Town Preservation
Act_ of 1966 covers a small number of places which
have played an important part in the cultural and
political history of the country, of which the most
famous are Kyoto, Nara and Kamakura. Areas in these
cities which are covered by historic buildings, and
the land adjacent to them, are being purchased by
the prefectural governments; and new developments,
including the felling of woodland, within them are
restricted. A quarter of the city of Kyoto now lies
within either Historic Landscape Conservation or
Scenic Beauty Districts under this and earlier
legislation, together with some of the forest-clad
mountains around it (Kaneyasu 1981, p. 47).
     However, zoning throughout the built-up areas
was not introduced until the _City Planning Act_ of
1968. This Act requires all cities and towns to
draw up a development plan for both the built-up
area and the surrounding land; and cities with
populations of 100,000 or more must include both
Urbanisation Promotion and Control Zones. The
delimitation of the City Planning Areas is the
responsibility of the prefectural governor, and he
may also indicate the general outlines of the plan
for the Promotion Zone. However, the details of
zoning are usually a matter for the city mayor and
council, although they must be reported to the
governor. In 1979 there were 1,858 City Planning
Areas in Japan, with a total population of a hundred
million people. Twenty-five percent of the country
lay within them, and three percent of this was in
the Promotion Zones. Nine percent was in the
Control Zones (Prime Minister's Office 1981, pp.
4-5). Developments in the Control Zones, other than
in existing residential areas, are not usually
permitted, and in the Promotion Zones those in
excess of a thousand square metres require

Table 5.1: Land Use Zones in City Planning Areas in Japan

Total area of City Planning Areas	9190 km^2
in which	
Urbanisation Promotion Zones	1578
in which land zoned for	
First type exclusive residential	326
Second type exclusive residential	290
Residential	480
Neighbourhood commercial	47
Commercial	63
Quasi-industrial	162
Industrial	87
Exclusive industrial	123

Source: Prime Minister's Office, Statistics Bureau, Japan
Statistical Yearbook 1981, Tokyo 1981, pp. 4-5.

permission. Promotion Zones and the built-up area
may also be divided into eight non-overlapping
sub-zones, each of which is designated for a
particular type of land use with which any new
construction ·must comply. The types are shown in
Table 5.1, and range from "First Type Exclusive
Residential", which is of relatively low-density
housing, through areas of higher density and
apartment block housing to commercial and industrial
areas. The total area in all the Promotion Zones
which lies within each sub-zone is also given in the
table, and from this it may be seen that about
seventy percent of the land is earmarked for
residential developments of one kind or another,
while most of the rest is intended for industry.
The types of activity which are prohibited in each
sub-zone are shown in Table 5.2; and each sub-zone
is also subject to restrictions on the density and
height of buildings. Some of the land in each of

Table 5.2: Restrictions on the Uses of Land within Urbanisation Promotion Zones

	Category 1 exclusive residential district	Category 2 exclusive residential district	Residential district	Neighborhood commercial district	Commercial district	Quasi-industrial district	Industrial district	Exclusive Industrial district
houses								▨
kindergarten, grade school junior high school senior high school							▨	▨
library, museum								▨
shrine, temple, church								
asylum for the aged, day nursery clinic								
college or university, senior professional school	▨						▨	▨
hospital	▨						▨	▨
retail stores (including department stores), restaurants	▨							▨
offices	▨							▨
hotel, motel, ryokan	▨	▨					▨	▨
bowling place, skating rink, swimming pool	▨	▨					▨	▨
theatre, movie theatre, entertainment hall, auditorium	▨	▨	▨				▨	▨
Japanese restaurant, bar, cabaret, dancing hall	▨	▨	▨				▨	▨
business warehouse, garage having more than 50m² floor spaces	▨	▨	▨					
small-scale food processing factory such as confectionary	▨	▨						
factory with the floor space of less than 50m	▨	▨						
factory with the floor space of less than 150m²	▨	▨	▨					
factory with the floor space of more than 150m²	▨	▨	▨	▨	▨			
factory considered to be dangerous or to deteriorate the environment	▨	▨	▨	▨	▨	▨		
storage of dangerous things such as gunpowder, petroleum, gas, etc.	▨	▨	▨	▨	▨	▨		
wholesale market, slaughter-house, crematory, sanitation facilities, disposal place	▨	▨	▨	▨	▨	▨		

☐ Building permitted

▨ Building prohibited

the sub-zones is, of course, already built on, but
there are also substantial areas awaiting
development which amounted in 1979 to almost 200,000
hectares, or 0.5 percent of the country.

THE COORDINATION OF CONTROLS

       In order to be effective such a system of
land-use control needs to be coordinated in at least
two regards. Firstly, in a situation in which each
type of zone is authorised by a separate piece of
legislation - Agricultural Zones by the Agricultural
Promotion Act, Urban Zones by the City Planning Act,
and Forest, Natural Park and Nature Conservation
Zones by the Forest, Natural Park and Natural
Environmental Conservation Acts respectively - some
overall plan of how each should develop in relation
to the others is required if a satisfactory balance
between land uses is to be achieved. Such a
balancing mechanism has been provided since 1974 by
the National Land Use Planning Act. This Act was a
response to the rapid rise in land prices in Japan
in 1972 and 1973, and to the increases in world
cereal and oil prices. It contained two major
groups of provisions - those empowering prefectural
governors to intervene in land transactions to
control land prices, and those requiring the
construction of a National Land Use Plan. However,
the boom in land prices had already passed before
the legislation was enacted, so the sections dealing
with this problem have never been implemented. On
the other hand, a National Land Use Plan has been
drawn up (National Land Agency, 1978).
       The Plan is based upon several objectives. It
is intended to increase the degree of
self-sufficiency in food production through the
increased use of double cropping, to make more
intensive use of the forests for both timber
production and recreation, to preserve the most
valuable of the natural habitats, to expand the
residential area, through the increased use of
multi-storey apartment blocks, and to provide
adequate land for industry and public facilities.
The expansion of the built-up areas is to be limited
to the Urbanisation Promotion Zones, and to further
reclamation of shallow, offshore areas in order to
provide more sites for industry. Preservation of
the environment is accorded high priority, as is the
provision of open space in urban areas and green
belts; and industry is to be encouraged to move out

Table 5.3: Land Use in Japan in 1972 and Targets for 1985 (in thousands of hectares)

	National			Metropolitan			Other Areas		
	1972	1979	1985	1972	1979	1985	1972	1979	1985
Agriculture	5990	5640	6110	750	660	630	5240	4980	5480
Cultivated land	5730	5470	5850	740	660	620	4990	4810	5320
Grazing land	260	170	260	10	–	10	250	170	250
Forest	25230	25280	24820	2090	2050	2060	23140	23230	22760
Natural Grassland	560	360	260	10	10	10	550	350	250
Water	1120	1140	1170	140	130	150	980	1010	1020
Roads	910	1030	1120	160	190	180	750	840	940
Built-up areas	1110	1380	1480	360	430	490	750	950	990
Residential	880	1070	1140	280	330	370	600	740	770
Industry	130	150	200	50	50	70	80	100	130
Offices and shops	100	160	140	30	50	50	70	110	90
(Urban areas)	640	N.A.	1160	330	N.A.	570	310	N.A.	590)
Other land	2820	2940	2820	400	450	400	2420	2490	2420
Total	37740	37770	37780	3910	3920	3920	33830	33850	33860

Sources: National Land Agency, National Land Use Plan (National Plan), Tokyo, 1978, p. 12; and

National Land Agency, The Outline of Annual Report on the National Land Use, Tokyo, 1981, p. 7.

of Densely-Inhabited Districts, thus eliminating
areas of mixed industrial and residential
development. However, the Plan does not envisage a
complete halt to the conversion of agricultural land
to other uses, and it makes no specific proposals
for the control of one of the most severely-affected
regions during earlier periods of industrial
development, namely, the coast.

The land-use targets in the Plan are for the
year 1985. They are based upon the recorded pattern
of use in 1972; and both sets of figures are given
in Table 5.3, together with the recorded pattern for
1979. It may be seen in the table that, under the
Plan, both the cultivated and built-up areas of the
country are expected to increase, but that there are
to be compensating declines in those of forest and
natural grasslands. The Plan also disaggregates the
targets into those for the three metropolitan
regions around Tokyo, Osaka and Nagoya, and for the
rest of the country. It shows that the net loss of
agricultural land in the metropolitan regions will
continue, but that it should be more than offset by
the conversion of forest and natural grassland to
agriculture in non-metropolitan Japan. It also
reveals that, in spite of the efforts by central
government to promote the relocation of industry,
there is expected to be considerable industrial and
other urban development in the metropolitan regions
at the expense of agriculture. However, it is
freely admitted in the Plan that, because the
forecasts of the allocation of population between
the metropolitan regions on the one hand, and the
rest of the country on the other, may turn out to be
wrong by two or three million, these regional
targets for land use may not be achieved.

All this may suggest that central government in
Japan can now exert effective control over land-use
change, but this is not so. In spite of the fact
that it is a ministry - the National Land Agency -
which draws up the National Plan, and the Prime
Minister who approves it, the work of designating
zones and enforcing them lies in the hands of the
forty-seven prefectural governors and the mayors of
the eleven largest cities. It is their
responsibility to incorporate the City Planning,
Agricultural, Forest, Natural Park and Nature
Conservation Zones into a prefectural or city Land
Use Master Plan, at a scale of 1:50,000, after
consultation with an appointed body of experts, the
Prefectural Council on Land Use Planning, and the
headmen of the municipalities. The Plan is then

submitted to central government for approval. Coordination of the prefectural plans with the National Plan is the responsibility of the National Land Agency; and a body of experts, the National Land Use Council, reviews the prefectural plans which are received. However, governors are not given targets for each land use based upon the National Plan before they begin to draw up the prefectural plans; the Land Use Council's review is not a formal process; and it has not been the practice of central government to amend the prefectural plans which are submitted, in contrast to the situation in Britain.

The second regard in which coordination is required in the system of land-use control is in respect of the substantial area of overlap between the various zones. Forty percent of the country falls into two zones, seven percent into three, and a very small area is common to no less than four; and any proposals for land-use change in such areas are subject to two, three or four sets of requirements. All the overlaps which could occur in theory are shown in Table 5.4, but several of them do not occur in practice, and the small areas which have been set aside for nature conservation pose only limited problems of zoning conflict. However, the problem is much greater for other zones, and the national government has laid down priorities to indicate which uses are to be preferred in such conflicts. For example, it may be seen that, in spite of the general presumption against urban development in the countryside at large, it is usually permitted where the site in question lies within an Urbanisation Control Zone or within a Forest or Agricultural Promotion Zone, but not where it is within a Reserved Forest or an Agricultural Land Zone; and forestry is usually given preference over agriculture, except where the site also falls into an Agricultural Land Zone.

Thus, there now exists a clearly-ordered structure of land-use controls and priorities which extends across the whole of Japan, but it is doubtful as to how effective it will be. For example, changes in land use between 1972 and 1979 (Table 5.3) indicate that it is extremely unlikely that the 1985 targets in the National Plan will be achieved. In particular, the agricultural area outwith the metropolitan regions was already well below target by 1979, and falling. Moreover, some urban uses, such as roads in the metropolitan regions, and offices and shops, had already exceeded their planned areas. Industry in the metropolitan

MAJOR ZONES / SUBDIVISIONS	Area — 1000 hectares	Area — Percent of country	Nature conservation — Ordinary district	Nature conservation — Special district	Nature conservation — Primeval nature environment preservation zone	Natural park — Other parts of natural parks	Natural park — Special zone	Forest — Other parts of the forest zone	Forest — Reserved forest	Agricultural — Other parts of the agricultural promotion zone	Agricultural — Agricultural land zone	Urban — Urbanization control zone	Urban — Urbanization promotion zone
Urban — Urbanization promotion zone	1553	4.1											×
Urban — Urbanization control zone	7525	20.2										×	×
Agricultural — Agricultural land zone	5524	14.8									×	×	×
Agricultural — Other parts of the agricultural promotion zone	12072	32.4								×	×	(1)	×
Forest — Reserved forest	7081	19							×	↓	×	↓	×
Forest — Other parts of the forest zone	18182	48.8						×	×	(5)	(4)	(3)	(2)
Natural park — Special zone	2975	8					×	○	○	↓	↓	↓	×
Natural park — Other parts of the natural parks	2174	6.1					×	○	○	○	○	○	(6)

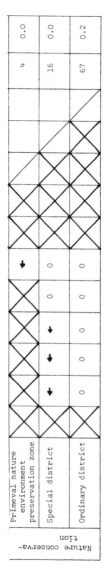

⊠ ↓

  : No significant overlap

○ : In case of overlap the land use indicated by the arrow is given priority.

↓ : In case of overlap land use change is allowed where it will make the uses compatible.

(1) : Urban use is permitted, as long as it is adjusted to the needs of agriculture.

(2) : Although priority is given as a rule to urban use green belts are protected.

(3) : Urban use is permitted as long as it is adjusted to the needs of forestry.

(4) : Priority is given as a rule to cultivation but forest use is permitted as long as it is adjusted to the needs of agriculture.

(5) : Priority is given to the forest use but agricultural use is permitted as long as it is adjusted to the needs of forestry.

(6) : Urban use is promoted but efforts are made to maintain the functions required for a natural park.

Table 5.4

General Guideline for Areas Zoned for more than one Use

Sources: National Land Agency, Land Use Master Plan, Tokyo, p.13; and, Rural Development Planning Commission, Rural Planning and Development in Japan, Tokyo 1981, p.52.

regions, in contrast, had remained constant in the
area of land it was using. Moreover, similar
failures have occurred within the cities, where
progress towards the segregation of non-conforming
activities has been slow; and there has been
frequent public opposition to proposals for the
redevelopment of congested areas.

Some of these failures stem directly from the
lack of power of the authorities to enforce their
plans. For instance, a study of the attitudes of
farmers in the Urbanisation Promotion Zones of
Tokyo, Osaka and Nagoya in the early 1980s revealed
that land for development may not be made available
in the places and to the extent which plans
envisage. Farmers in such Zones are well placed to
sell their land for development, but thirty-seven
percent of those questioned indicated that they
intended to continue to farm it, while forty-seven
percent were willing to sell only a part of it, and
only sixteen percent were prepared to sell it all
(National Land Agency 1981, p. 38). Failure to
release land on a larger scale or in a more orderly
manner than this will have at least two consequences
for land-use patterns. It will be necessary to
extend the Promotion Zones more quickly than would
otherwise have been required in order to continue to
provide ten years' supply of developable land; and,
in a country in which almost all farmers hold their
land in at least four or five small and scattered
parcels, the limited and partial release of those
holdings is likely to give rise to further piecemeal
and sporadic development in the urban fringe, after
the manner of the 1950s and 1960s. The response of
the National Land Agency to this situation in its
1982 <u>White Paper on National Land Use</u> was to argue
for an expansion of the Urbanisation Promotion
Zones, and a. reform to encourage farmers in them to
sell land to developers (pp. 21-22).

Plans may also not be carried into effect
because local authorities have it in their power to
frustrate developers, even where the proposed
development is in accord with the Land Use Plan.
For example, during the 1950s and 1960s prefectures
vied with each other to attract industry (Sargent
1980, pp. 207-208) with little regard for the
regional policies of central government; and in the
1970s municipalities near to large centres of
employment placed severe restrictions on housing
developments, and charged developers for the
provision of such public amenities as school
buildings, in cases where the occupants of the

houses were not likely to be working in the city,
but to be commuting elsewhere (Hayakawa 1978, pp.
25-26). Similarly, some have acted ahead or against
national policy, and halted plans to build new
industries in controversial locations (Sargent 1980,
p. 210).

THE NATIONAL CAPITAL REGION

Nowhere have the pressures upon land been
greater in Japan during recent decades than on the
Kanto Plain (Figure 5.4). The largest area of flat
and low-lying land in the country has been the site
of its most populous urban settlement, and one of
the world's largest cities, since the eighteenth
century; and, since the Second World War,
substantial industrial and commercial growth has
occurred in Tokyo, its outport Yokohama, and in
other towns within the conurbation. Over this
period the concentration of the nation's secondary
and tertiary activities in the National Capital
Region has increased considerably, so that, for
example, about half of the headquarters of the
publicly-quoted businesses in the country and of the
places in tertiary education are to be found in the
southern part of the Kanto Plain (National Land
Agency 1977, pp. 3-4, 19-20). This rapid economic
growth has attracted several million people into the
Tokyo commuting zone, especially from parts of the
country with lower levels of earnings. Between 1955
and 1965 net immigration to the Region exceeded the
natural increase of the population there; and,
although the net inflow fell sharply to very low
levels in the late 1970s, it was still positive.
Meanwhile, natural increase alone added 1,500,000 to
the population between 1975 and 1980. The
consequence of all this has been that the population
of southern Kanto (Tokyo, Chiba, Kanagawa and
Saitama prefectures) has risen from fifteen to
twenty-nine million between 1955 and 1980, and is
still growing (Prime Minister's Office 1981), while
the tide of building has spread far beyond the
boundary of Tokyo prefecture. However, population
has been declining since the early 1960s in an
every-widening band around and including the centre
of Tokyo, as commercial activities have outbid
residential land use, and as the inhabitants have
sought larger dwellings than those which they could
afford in central locations. The strength of this
competition for land has been reflected in its

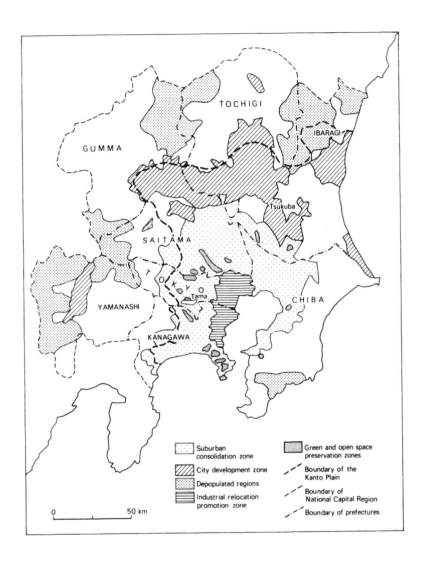

Figure 5.4:   The National Capital Region of Japan

price,   which   has risen to extraordinary levels in
the immediate environs of Tokyo (Table 5.5),   and in
the continuing reclamation of large sites along the
edges   of   Tokyo   Bay   for   industry   and   port
facilities.
    These   changes are illustrated in Table 5.5 for
three of the prefectures within the National Capital
Region -   Tokyo,   the   'suburban'   prefecture   of
Saitama,   and Yamanashi,   which is situated in   the
mountains of central Honshu, and, unlike the others,
experienced   population   decline   between   1950   and
1970, and a net out-migration since then.   The table
also shows how land use has changed since 1960.   For
example,   the areas under dry fields and paddy have
fallen   in all three   prefectures,   although   to   a
markedly   smaller   extent   in   Yamanashi than in the
others;   and the forested area has increased   in
Yamanashi,   while it has declined in the other two.
However,   in spite of the large   increases   in   the
residential areas in Saitama and Tokyo, the quantity
of housing available per capita still remains   much
lower   in   Tokyo and its suburbs than in the largely
rural prefecture of Yamanashi.
    Government   reaction to these changes has   been
weak.   The first National Capital Region Development
Plan of 1956 identified three major zones: an inner
one,   comprising the twenty-three wards of the city
of Tokyo and the neighbouring towns of Mitaka and
Musashino,   which has been subject to a ban on any
major new industries, colleges or universities since
1959 - a ban which has been extended subsequently to
cover   Kawasaki and Yokohama;   a Green Belt of about
ten kilometres width with an inner boundary of about
fifteen   kilometres   from   the   city   centre;   and   a
peripheral   zone   beyond   the   Green   Belt   in   which
satellite   towns   were to be built.   The   Plan   drew
heavily for its proposals upon Abercrombie's Greater
London Plan.   However,   control over the height of
buildings   was   relaxed   in the 1960s,   thus allowing
high-rise   developments   in   the   city   centre   and
Shinjuku,   which increased the intensity of land use
there;   expansion of existing   industry   was   still
allowed,   including   new   factories   of   up to five
thousand square metres, and so the number of jobs in
the city continued to grow;   and the Green Belt was
abandoned in 1965 as a failure, to be replaced later
by   much   smaller protected Green Areas.   One of the
chief achievements of the Plan was the establishment
of Tama New Town beyond the edge of the built-up
area to the west of Tokyo,   but,   although it
contained some industry, many of the people living

Table 5.5: Some Statistics of Selected Prefectures

| | Prefecture | | |
	Tokyo	Saitama	Yamanashi
Area (square kilometres)	2156	3799	4463
Area under various landforms (percent) –			
Mountain, volcano and hill land	52	38	88
Lowland and upland (plateaux)	48	62	12
Population (thousands)			
1950	6278	2147	812
1970	11408	3867	762
1979	11596	5309	795
The changing use of privately-owned land 1950-1979 (percent)–			
Paddy	-82	-17	-11
Dry fields	-47	-24	- 5
Forests	- 2	- 6	+18
Residential	+47	+129	+110
Price of land in 1980 (thousand yen per 0.1 hectare) –			
Paddy	66869	958	873
Dry field	83105	770	699
Residential	197000	78600	28600
Commercial	636400	201600	82600
Housing space in 1980 (tatami per person: two tatami cover 3.3 square metres)			
	7.1	7.4	9.1
Area covered by city planning (square kilometres)			
	1726	2621	778
in which Urbanisation Promotion Zones	1086	665	77
Reserved forest	137	436	1846

Source: Prime Minister's Office, Statistics Bureau, Japan
Statistical Yearbook 1981, Tokyo 1981.

there work in Tokyo. In fact, urbanisation in the
late 1950s and 1960s proceeded outwards from the
Tokyo conurbation in almost all directions, so that
the clearly-identifiable and segregated land-use
areas envisaged in the _Plan_ did not come into
existence.

Subsequent revisions have created a plan which
includes an Industrial Relocation Zone, which covers
the twenty-three wards and four cities from which
major new industry is still excluded; a Suburban
Consolidation Zone in the area around the central
cities, and in which the land is not yet fully built
up; and City Development Zones beyond that, which
are intended to absorb most of the urban growth in
the future. Tax incentives have been instituted to
attract industry from the Relocation Zone to
industrial parks in the other two Zones (National
Land Agency 1977, p. 8). One of the Development
Zones is at Tsukuba, in the north-east of the Kanto
Plain. The National Government began to buy land
there in 1966, and to encourage the movement of
institutions of higher education and research from
Tokyo. By 1980 the population of the new town had
reached 120,000, and forty-three establishments had
relocated there. However, little privately-owned
industry had been attracted, and the population was
80,000 short of the target (National Land Agency
1980).

In addition to these plans by central
government the Kanto Plain is also covered by the
prefectural planning system, which was established
between 1968 and 1974; but comparisons of the
National Capital Region Development Plan in its
present form with prefectural land-use plans reveal
some remarkable discrepancies. For example, most of
Tokyo prefecture outwith the Industrial Relocation
Zone falls within the Suburban Consolidation Zone,
thus indicating that it is an acceptable area in the
view of central government for further development.
However, this zone includes all the Urbanisation
Control Zone of the prefectural plan. Such a direct
contradiction suggests that the Capital Region Plan
is no more than a very broad and generalised pattern
for future development, while the prefectural plan
is based upon a more detailed identification of the
buildable land. For instance, most of the remaining
cultivated land in the prefecture has been placed in
the Urbanisation Promotion Zone in the prefectural
plan (Table 5.6), and the steeply-sloping land,
which is forested at present and would probably be
unattractive to developers, has been assigned to the

Table 5.6:   Tokyo Prefecture - Zoned and Cultivated Land in
             1978

	Percent of Tokyo Prefecture	Of which Cultivated
Urbanisation Promotion Zone	48.6	4.2
Urbanisation Control Zone	17.9	1.4
Unzoned City Planning Area	14	0.9
Outwith the City Planning Area	19.5	0.2

Source: Tokyo Metropolitan Government, <u>Tokyo Tomorrow</u>,
          Municipal Library No. 17, Tokyo 1982, p. 31.

Control   Zone.   However,   in Saitama prefecture   the
boundary   of   the   Consolidation   Zone has little in
common   with   either   those   of   the   Urbanisation
Promotion   or   Control   Zones,   so that national and
prefectural   planning appear to be in   even   greater
conflict there.
     A   more   detailed   picture   of the incidence of
land-use   control is given for   a   part   of   Saitama
prefecture   in   Figures   5.5   and   5.6.   These snow,
respectively,   some of the chief land uses, and some
of   the   land-use   zones   in   an   area of thirty-six
square   kilometres on the northern outskirts of   the
Tokyo   conurbation.   Several   major   elements can be
identified   from the land-use   map.   Firstly,   the
Arakawa river,   which crosses the area,   is enclosed
within huge artificial levees,   and the land between
them   is   given   over   almost entirely to uses which
will   not be damaged by periodic flooding,   such   as
playing fields and golf courses.   Secondly,   much of
the   land beyond the levees   is   also   part   of   the
natural floodplain of the river and its tributaries,
and is in use as paddy.   Sinuous bands of settlement
occur in this area on the former,   natural levees of
the rivers, which are only slightly raised above the
surrounding   rice   fields.   Lastly,   the   city   of
Kamifukuoka   lies   on   a   low   terrace   above   the
floodplain.   Most   of   the   land   in   it   is   in

residential use, but there is also much dry-field farmland in small and isolated parcels in the northern part, some of which has already fallen out of agricultural use, and is idle; and there are also some large industrial sites and commercial uses in the city.

Some of the zoning of this area is shown in Figure 5.6. At the national level the whole area lies within the Suburban Consolidation Zone of the National Capital Region Development Plan, except for the enclosed part of the Arakawa floodplain, which is designated as a green or open space. However, at the level of the prefectural plan it is clearly not intended that the remaining agricultural areas between Kamifukuoka and the cities of Kawagoe to the north and Omiya to the east should be built over in the near future, for the entire area lies with the City Planning Areas of the surrounding towns, and the boundaries of their Urbanisation Control Zones closely follow the edge of the floodplain for the most part, restricting development to the plateau and to some, but by no means all, of the levee areas. They are also the boundaries of the Agricultural Promotion Zone. However, the Agricultural Land Zone is of more limited extent. Much, though not all, of the area of rural settlement is excluded from it, but so also is almost all the plateau, dry-field land, and some of the paddy. Within the Urbanisation Promotion Zone only six of the eight sub-zones are represented, but it may be seen that almost all the small, isolated parcels of dry-field land are zoned for residential development within the next ten years. In contrast, only small parcels of paddy on the edge of the built-up area are to be developed. Elsewhere in the urban area the pattern of zoning follows the existing division between commercial, industrial and residential land uses very closely. Thus, if the prefectural plan for this area is carried out, and suburban growth in the Tokyo conurbation continues, the present areas of settlement - both urban and rural - will become more compact, while most of the paddy will remain in cultivation. However, there does not appear to be any intention, at least in this area, of enforcing a greater segregation of nonconforming land uses in the built-up area.

CONCLUSION

Japan has adopted land-use regulation as a means of dealing with its land problems relatively

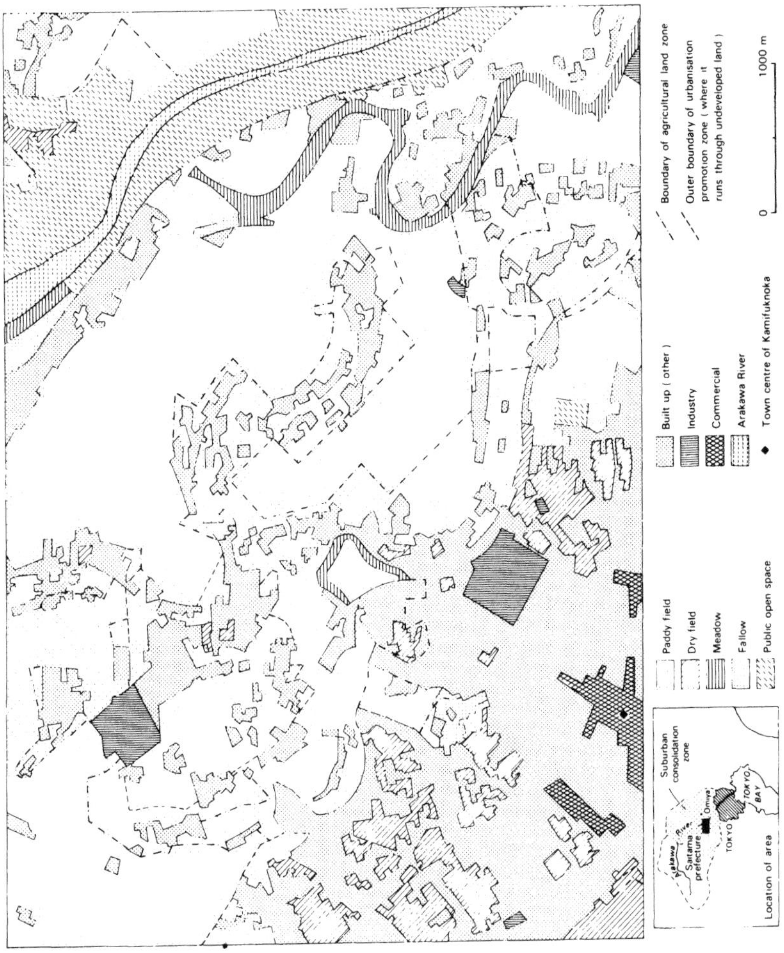

Figure 5.5:

Land Use in the

Northern Fringe of

the Tokyo Conurbation

in 1981

Figure 5.6:

Land-Use Zoning in the

Northern Fringe of the

Tokyo Conurbation in

1981

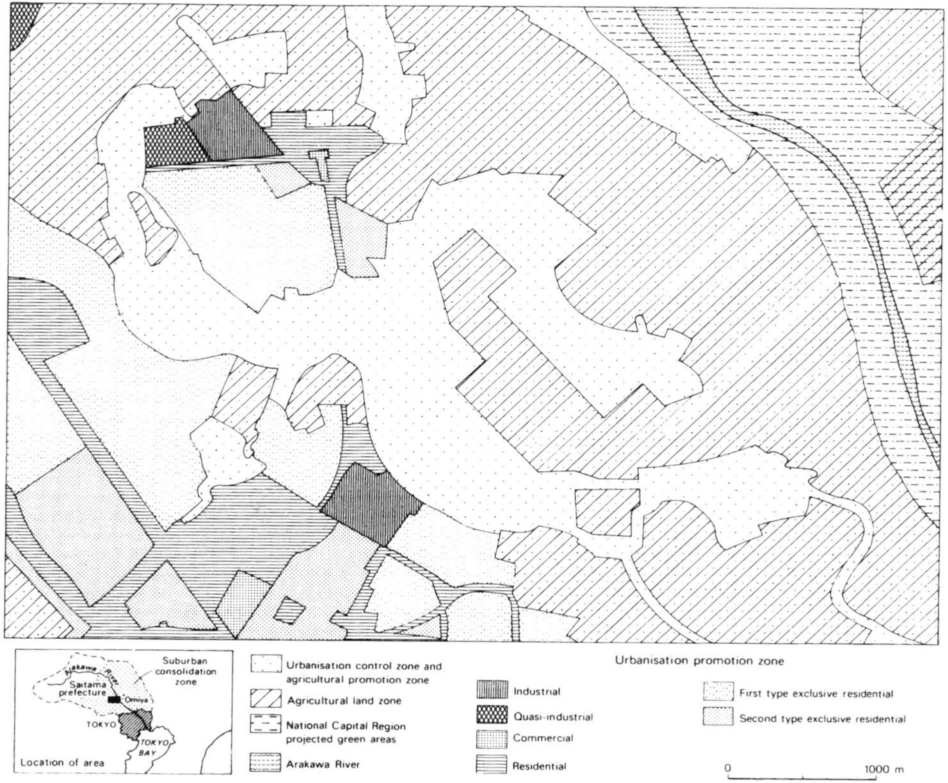

recently;   but having done so it appears,   at least
from   the legislation,   to have   created   a   uniform
structure   of control across the whole country,   and
one   which links   the   tiers   of   government   in   an
orderly   and   hierarchical   manner.   However,   this
appearance   is   illusory.   The   power   of   central
government,   though   substantial,   is not nearly as
great as the legislation suggests; and the scope for
variety   in land-use policy at the prefectural level
is considerable.   Of course, major areas of decision
making   with   important   implications for the pattern
of   land   use   are   the   prerogative   of   the   Prime
Minister   and   the   Cabinet.   They   include   the
designation   of Natural Parks,   the   management   of
state   forests,   and   the   choice   of sites for new
towns,   airports,   large   reservoirs   and   the
shinkhansen   railway   routes.   Government   has also
acted   to protect the population   from   the   hazards
which   are   latent   in   a   country   of steep slopes,
torrential   rainfall and earthquakes by   restricting
the   pattern   and   intensity of development;   it has
direct   control   over   the   transfer   of   all   large
parcels   of   land   out   of   agriculture;   and,   by
manipulating   the prices for agricultural   products,
and   especially   for   rice,   it has discouraged the
release   of farmland for development and raised land
prices.   However, it has met considerable opposition
since   the   late   1960s   to   a   number   of   major
investments,   and   has failed to achieve several of
its   plans   for   the   spatial   development   of   the
economy.
     It could be argued that,   in spite of the great
pressure   of demand for land,   Japan does   not   need
detailed   controls.   Economic growth since the 1950s
has   banished the   spectres   of   overpopulation   and
hunger   which   had   plagued   the   country before the
Second World War, and has provided a rising standard
of   living   during a period when the cultivated area
has   been   declining.   Moreover,   the   increasing
dependence   upon   imports   of   both   farm and forest
products   since the war - Japan was   importing   more
farm   goods   than   any other country in the world by
the   end of the 1970s - need not necessarily be   the
strategic risk which it may appear to be,   for there
is   considerable   scope   for   a   return   to   double
cropping on farms,   and Berque (1980,   pp.   158-162)
has   claimed that the productivity   of   the   forests
could be raised substantially if the present pattern
of   fragmented ownership of forest land were   to   be
altered.   Thirdly,   it is quite plain that the very
high   prices   for   land   inhibit   changes   of   both

ownership and use. Not only do farmers cling to their land when they could sell it, but local government has complained that its price is such as to reduce significantly the amount of public investment in housing and other land-using amenities which it can provide (Tokyo Metropolitan Government 1976, pp. 8-16). It would seem unlikely therefore, that land would be used profligately, wastefully, or in a manner which would lead to its deterioration.

Yet this is exactly what has happened. In the late 1950s and 1960s spillover effects from industrial and urban growth polluted the atmosphere, the rivers and the sea, and led, through the extraction of groundwater, to the subsidence of land in areas of Tokyo which were already in floodable locations (Tokyo Metropolitan Research Centre 1979, pp. 57, 88, 96). These actions suggest that, in an economy in which one of the factors of production is as expensive, or more so, than in competing countries, the incentive for producers to avoid some of that cost by externalising it is considerable; and that the need for government to enforce the internalisation of that cost, or to reduce its incidence upon the general public by segregating nonconforming land uses, is equally great. Secondly, it can be argued that, even in the absence of direct intervention by the state to plan land use, the political relations within Japan have been such as to force the allocation of land away from that which a truly free market would have achieved, and that to a very considerable extent. The weakness of the national government in the face of farmers' demands for guaranteed prices and import controls has led to a situation in which the area of farmland, and therefore the price of land for other uses, and especially housing and public amenities, are both grossly excessive. In this respect the Japanese have failed to carry through the policy which has otherwise characterised their post-war economic management - of encouraging the transfer of factors of production from less to more productive uses - and therefore it may not be correct to claim, as Hall (1977, p. 238) does in the quotation at the start of this chapter that, in the pursuit of rapid economic growth, there has been an "overwhelming desire to exploit every opportunity" in Japan. If that is to occur with respect to land, the degree of government intervention in agricultural markets must not be increased - it must decline.

Chapter 6

LAND-USE REGULATION IN THE UNITED STATES OF AMERICA

"Land use decisions today are inconsistent, irrational, inequitable, and inefficient. They are frequently unconstitutional as well. There is no common national referent for land use decisions; there are no agreed-upon land use goals or standards. The present fractionated method of decision making is extremely costly. It has led to a squandering of our resources, and has still failed to provide the "decent home and suitable living environment for every American family" that was promised by Congress long ago in 1949.
Why do we cling to this system?"
(A.L. Strong, Land as a Public Good, in *The Land Use Policy Debate in the United States*, ed. J.I. de Neufville, New York 1981, pp. 217-218).

"It would appear that American land use is already overcontrolled." (R.H. Jackson, *Land Use in America*, London 1981, p. 4).

INTRODUCTION

Unlike Britain and Japan the United States of America covers a very large area and enjoys a rather low density of population (Table 1.2). It also possesses a system of government which allows for considerable variety of action among members of the lower tiers. Nevertheless there is a strong and recognisable land ethic among the population which has influenced, and perhaps restrained, the development of land-use controls, at least until the 1960s. It is difficult to do justice both to the central themes and to the wide variety of action at

federal, state and local levels in such a large
country in a single chapter, but an attempt will be
made to indicate how the attitude of government
towards land use has changed since Independence, as
perceptions of the supply of land have altered, to
describe some of the most important of the
participants, to outline the most pervasive and
detailed of the systems of local regulation -
zoning - and to assess the expanding role of the
federal government in land-use control during this
century. The chapter will conclude with an
examination of the policies and legislation at the
level of the individual states, and a description of
the combined incidence of controls by the different
levels of government in three states which have not
featured widely in the literature on this topic
before: Massachusetts, Minnesota and Arkansas.

DEMANDS FOR LAND AND ATTITUDES TOWARDS IT SINCE
INDEPENDENCE

     For much of its history the United States has
seemed to possess enormous and inexhaustible
supplies of land; and American society has
traditionally accorded a very high priority to
private ownership, and to the freedom of owners to
do with their land as they wish. Indeed, the
availability of land, and the possibility of owning
the freehold to it, have been at the basis of the
development of the American nation, for many
Americans are the descendants of immigrants who fled
across the Atlantic precisely in order to take
advantage of these conditions. Many of the original
settlers came from European countries in which land
ownership was feudal in character, or in which land
was usually held in tenancies on large estates, and
where feudal superiors and landlords behaved in an
autocratic manner; and one of the chief purposes for
emigration was to acquire the wherewithal in the
form of cheap land to be both independent and free.
As a result, most immigrants to the States welcomed
such statements as that in the Massachusetts Bill of
Rights that

     "all men... have certain natural, essential and
     inalienable rights, among which [is] that of
     acquiring, possessing and protecting property",

and believed that any regulation of such matters by
the state was a threat to their independence and to
democracy. These views are widespread in the United

States to this day, and have been supported over the years by the courts, which have held that the imposition of controls over the use of privately-owned land without compensation can represent a 'taking' of some of its value, and as such is in violation of the Fifth Amendment of the Federal Constitution.

This attitude found clear expression in the land policy of the federal government during the first hundred years of the United States' history. High priority was given to the acquisition and orderly settlement of territory, especially where its ownership might be disputed by other countries; and huge tracts were added to the thirteen original states by the Louisiana Purchase of 1803, the acquisition of Florida in 1819, of Texas in 1845, of the Northwest in 1846, and of the Southwest in 1848 and 1853. However, this land was speedily disposed of to settlers through the Ordinances of 1785 to 1832, the Pre-emption Act of 1841 and the Homestead Act of 1862. Much was sold to raise capital for the federal government; much was granted to railway companies to subsidise the construction of transcontinental lines; much was given to endow colleges; and enormous areas were sold at minimal prices to individuals, in the Jeffersonian belief that the best society is made up of independent, yeoman farmers. So successful was this policy of land disposal that the proportion remaining in federal hands in the eastern and central two-thirds of the country does not exceed eight percent in any state, with the exception of New Hampshire, where it is twelve, and South Dakota, with eighteen. Only in the West have large areas remained in federal ownership, varying from thirty-five percent in Washington through forty-five in California to seventy-three in Arizona and Utah, and a peak of eighty-five percent in Nevada, so that federal agencies own more than a third of every western state.

Thus, the federal government pursued its policy with regard to land ownership with some vigour, but it paid little or no attention to the subsequent use of the land. Indeed, the thirteen original states never relinquished their right to control land use to the federal government; the Ordinances of 1785 and 1887 guaranteed non-interference by the federal authorities in any subsequent sale or use; and the regulation of the use of privately-owned land was delegated to the states in general in the Tenth Amendment to the Constitution in 1791. Federal interest in land-use control only developed towards

Table 6.1: Some Characteristics of Selected States

	Arkansas	Massachusetts	Minnesota	U.S.A.
Area (square kilometres)	134361	21303	216895	5860065
Population in 1980 (thousands)	2286	5737	4076	222159
Increase in population 1950-80 (percent)	19.7	22.3	36.7	47.4
Density of population in 1980 (per square kilometre)	114	1793	124	253
Land in federal ownership in 1980 (percentage of total)	9.9	1.5	6.6	33.9
Area of farms (thousand hectares)				
1930	6497	811	12511	399346
1950	7837	672	13308	470023
1969	6352	284	11674	430337
1978	6304	276	11700	417205
Percent of state	46.7	12.9	53.9	71.2

Area of commercial timberland (thousand hectares)

1953	7807	1319	7324	214556
1970	7368	1413	6829	202227
1977	7368	1132	5542	195268
Percent of state	54.6	52.9	25.5	33.2
Property taxes as a proportion of all revenue of municipalities and townships from own sources in 1970 (percent)	25.8	50.3	38.7	39.3

Source: United States Department of Commerce, Bureau of the Census, Statistical Abstract of the United States 1972, 1981, Washington, D.C.

the end of the nineteenth century, and then only
with reference to a few, limited parts of the area
remaining in federal ownership. Yosemite Park was
set aside for the public on grounds of scenic
interest in 1864, Yellowstone Park in 1872, and the
Adirondack Park in 1885. On the other hand, the
Mining Law of 1875 confered the right to extract
minerals from public land on anybody who had filed a
claim; and up to 1890 it was generally agreed that
there was little or no justification for government
at any level to exercise control over land use.

But there has been a considerable change since
then. Under the <u>Forest Reserve Act</u> of 1891 powers
were given to the President to withdraw forest land
from public entry, and to set up forest reserves,
with the effect that by 1909 141 million acres had
been designated, chiefly in the West; a National
Park System was established in 1916, and all other
federal land was withdrawn from homesteading in 1934
until it could be classified in terms of grazing
quality and suitability for agriculture. Thus the
frontier was closed, and the extension of the
oecumene gave way to the intensification of its
use. It is true that much land remained in federal
ownership, and that large amounts are still
available for settlement, lumbering and mining, but
it is land which the pioneers had rejected as
inferior as the frontier moved westward during the
nineteenth century. No longer are there supplies of
well-watered and fertile land waiting for the
younger sons of farmers in the east or for
immigrants from Europe; and in the twentieth century
fortunes, or at least personal independence, have
had to be won by the improvement of what is already
settled.

What is more, the area which has been available
for such improvement appears to have been
declining. As population has grown (Table 6.1)
farmland has fallen, and the area of prime
agricultural land was reduced by more than a million
acres, or 0.4 percent per annum, between 1967 and
1975. The area of commercial timberland also
declined considerably after the 1950s; and much
coastal, forest, lake-shore and mountain land became
the site of suburban, ex-urban and second-home
development, tourism and recreational activity.
Wetland and wilderness diminished; and vast areas of
rural land were strip-mined for coal, paved over for
roads or covered by fast-food restaurants, petrol
and power stations. Moreover, much of the land
which has been taken out of agriculture has not been

put to any other use, but lies idle; and much land
and property in the cities was abandoned in the
1960s and 1970s, and lies unused and derelict.
Thus, what were formerly thought to be inexhaustible
land resources have been subject to a wide range of
recent demands.

Nor is it likely that these demands will be
significantly reduced in the future. An estimate in
1972 suggested that twice as much recreational land
will be required in 1990 as was used in 1960, a
third more will be needed for transport, a fifth
more for public installations and facilities, and a
tenth more for commercial forest, as both population
and income levels continue to rise (Healy 1979, pp.
17-20). Thus, it is not surprising that there was
an upsurge of public concern in the late 1960s and
early 1970s, which was epitomised by 'Earth Day' in
1970, and a growing, but by no means universal,
acceptance of the need for some government control
over land-use change and the introduction of new
forms of regulation. Several authors have claimed
that land is not now perceived to be so much a
commodity to be bought, sold, used or employed for
speculation, as it was in the past, as a resource to
be protected and conserved; and two, F.P. Bosselman
and D. Callies, have identified what they have
termed The Quiet Revolution in Land Use Controls
since the mid-1960s as a consequence of this shift
in perception. However, no general policy has
emerged for the nation as a whole; and the pace and
extent of the adoption of controls have varied
widely across the country.

In view of the statistical evidence which has
been mentioned above about the changing pattern of
land use in the United States, it may be something
of a surprise to learn that one reason for the lack
of a consensus in favour of controls - in contrast
to the situation in Britain after the Second World
War - has been the difficulty in establishing
exactly what has been happening to land in America
during recent decades. The federal Department of
Agriculture holds censuses every quinquennium which
yield information about the land on farms; the Soil
Conservation Service started the National Resources
Inventory in 1977, which covers rural land which is
not in federal ownership (Berry 1980, pp. 159-160);
and the area of commercial forest is also reported
periodically; but there is no nationwide measurement
of some of the most controversial results of
land-use change in recent times, namely the
expansion of the built-up area and of derelict and
idle land. Nor have the many local studies (Berry

1978, Brown et al. 1977, Coughlin 1979, Coughlin
et al. 1977, Furuseth 1979, Ziemetz et al. 1976)
adequately filled this gap, for they have tended to
concentrate on the urban fringe, and thus have
ignored evidence from the huge rural areas.
Moreover, they are not comparable with each other,
and frequently cover very brief periods which may
not be typical of, say, the long-term investment or
building cycles of which they are a part. One of
the most substantial is that by McConnell (1975)
which attempts to give a comprehensive picture of
land-use change in Massachusetts between 1952 and
1972 from the evidence of aerial photographs. Even
so, it covers only a small proportion of the
country, was extremely time-consuming to carry out,
and has not been repeated in comparable form for
other states. In the absence of nationally-surveyed
information both government and private
investigators have fallen back on extrapolations,
based upon the small areas which have been mapped,
and estimates based upon the residual balance of
land after accounting for alterations in the areas
of farms and forest. However, Hart has drawn
attention to the fact that the estimates which have
been produced by these means of, say, the transfer
of land out of agriculture or of the rate of
development, have varied so widely as to be of
little value as a basis for decisions about the need
for land-use regulation or the direction which such
regulation should take (Hart 1976).
      A second reason for the continuation of the
original American attitude to land amongst a large
number of the population, in spite of the recent
demands upon land, has been the continuation of the
perception of the United States as a large country.
Even if some of the more pessimistic estimates of
the rate of land-use change are correct, it is clear
that the land resources which are available to the
nation are still very substantial. For instance, in
spite of all the urban growth which has occurred, it
is not expected that the urban area will exceed four
percent of the land surface by the end of the
century (Hart 1976); and it has been estimated that
the requirements for farm and rangeland will
decline. One estimate suggests that a quarter of
the country may be 'surplus' to requirements in the
1990s, and that there will be no general shortage of
land in the near future. However, there may be
increasing competition and dispute over particular
pieces of land, and the use to which they should be
put (Healy 1979, p. 20).

SOME AMERICAN LOBBIES

Of all the countries which are included in this study, the United States offers, perhaps, the clearest picture of the way in which those disputes arise and are resolved, for the registration of official lobbyists, and the relatively open style of government, permit the identification of the actors who are involved, and the policies for which they are arguing, with greater ease than in Britain, with its tradition of closed government, or Eastern Europe, with its habits of secrecy. "Our government is largely a brokerage for special interests", claims Ophuls (1977, p. 190); and indeed, the number and variety of pressure groups is very large. However, many pressures are also brought to bear in 'smoke-filled rooms'; and, in spite of the official openness of American government, corruption is thought to occur with some frequency.

As in both Britain and Japan the producer groups occupy a particularly important place in decisions about land, and amongst these the farming lobbies are of special interest. Guither (1980, pp. 12-13) has reported that no fewer than 611 different groups and individuals submitted their views to the agriculture committees of the United States Congress in 1977 alone, and that many of these represented farmers. The purpose of these farmers' groups is to obtain the best possible conditions for their members' activities, and in particular price supports, cash payments for the fallowing of land in order to reduce output, and export subsidies. At first sight it might appear that their influence would not be great. Not only is the country more than self-sufficient in the production of most foodstuffs, but there are also consumer and welfare lobbies whose opposition to high levels of agricultural support was growing during the 1970s (Lyons and Taylor 1981, pp. 129-130). However, the farming groups have great advantages. Firstly, they are large. For example, the National Grange - an organisation representing family farms which was founded in 1867 - has a membership of half-a-million in forty-one states; the National Cattlemen's Association has 280,000 members, and the National Association of Wheat Growers about eighty thousand. Secondly, they are wealthy. During the lobbying associated with the 1977 <u>Food and Agriculture Act</u> Associated Milk Producers, Inc. made contributions to the funds of no less than 222 congressional candidates, totalling

456,000 dollars, and the Mid-American Dairymen,
Inc. supported 197 to the extent of 229,000 dollars
(Guither 1980, pp. 325-327). Thirdly, Hathaway
(1981, pp. 779-780) has suggested that since the
1950s the agricultural committees in Congress have
become dominated by the representatives of the
agricultural commodity groups. Furthermore, the
population of the rural states is over-represented
in the Senate. Thus, for all these reasons, the
farmers have been very successful in their attempt
to influence federal policies. For instance, the
dairymen received such a high level of price support
in the 1977 Agriculture Act that huge surpluses of
dairy products accumulated in the hands of the
federal government during the early 1980s.
Similarly, the wheat growers were able to persuade
President Reagan to lift the embargo on grain
exports to the Soviet Union in the early 1980s
despite that country's continued military occupation
of Afghanistan; and overall, the farming sector
"enjoyed the largest net gain in real wealth of any
other major sector in the economy" in the 1970s
(Hathaway 1981, p. 779), with the effect that land
which would otherwise have fallen idle has been kept
in production, and wet and other marginal land,
which might have lain undisturbed, has been improved
and cultivated.

Among other producer groups which have strong
views about land-use regulation is that of the
developers. However, there is a marked contrast in
attitudes between the small and the large-scale
builders.

"The suburban builder usually views public
officials as gatekeepers who rarely help but
often hinder the development process" (Baerwald
1981, p. 354);

and one of their chief lobbyists, the National
Association of Home Builders, which had more than
fifty thousand members in the 1970s, opposed the
National Land Use Planning Bills, which failed to
pass in Congress at that time (Pynoos 1980, pp.
32-45). The big builders, in contrast, would
probably approve of stronger systems of control,
especially if these were to be achieved by the
replacement of the multiplicity of municipalities,
each with its own development regulations, by much
larger, metropolitan authorities, because this would
lead to standardised controls over large areas,
which would make the work of such developers simpler
and more profitable. It would also mean that they

would be faced with bodies which would be further from the public, and which would probably be less influenced by objections to building proposals from localised interest groups (Walker and Neiman 1981, pp. 67-83).

Producer groups such as these are frequently opposed by one or more of the very large number of amenity and conservation organisations which exist in the United States. Some of the most influential of these are the Audubon Society and the Nature Conservancy which, like the National Trust in England, own land for the purposes of conservation. Since its foundation in 1949 the Nature Conservancy has acquired more than 400,000 hectares of land, much of which it has handed over to government so that its scenic quality may be preserved (Irland 1979, pp. 26-27, 140). Another group - the Sierra Club - enjoys a membership of 170,000, and has had sufficient financial resources in the past to take the federal government to court in order to halt lumbering and mineral extraction in some roadless, wilderness areas. Both of these organisations have a large number of local branches through which they lobby government on land-use matters at all levels. They are deeply suspicious of such producer groups as the National Cattlemen's Association, the American Mining Congress, the lumber lobbies and the recreational groups, such as the American Rifle Association and the off-road-vehicle clubs, and also of the Bureau of Land Management and the Forest Service of the federal government, which they believe to have been 'captured' by the producer groups. Their high point may have been the passage of the National Environmental Protection Act of 1969. However, they have never enjoyed resources on the scale of those of the producers; and the level of donations to them has been falling since the mid-1970s (Ames 1981, pp. 9-14).

Culhane (1981) has reviewed the debate about the 'capture' of federal agencies with particular reference to both the Bureau and the Forest Service. These two agencies together are responsible for about a third of the area of the United States, much of which is also of interest to foresters, ranchers, the mining industry, and the general public for purposes of recreation, as well as to the conservationists. It has been suggested that 'capture' occurs for two reasons. Firstly, it is the local officers within such large organisations who are responsible for setting felling rates or grazing densities on public lands, but in doing so they are obliged to deal largely

with local producers, and may wish to avoid
persistent criticism of their decisions by such
groups. Secondly, where the congressional
committees which oversee the activities of these
agencies include large numbers of members from
states in which the agencies are major land-holders,
those committees are likely to be critical of
policies which are opposed by their constituents,
and to report adversely against them when they are
effective in limiting the activities of local
industries. However, a factor analysis of the
attitudes of the organisations studied by Culhane,
from which an environmental-utilitarian attitude
scale was constructed, revealed that the Bureau, the
Forest Service, and state fish and game agencies lie
mid-way between the conservationists, on the one
hand, and the lumber, stockmen and miners on the
other. It also showed that the fishing, hunting and
off-road-vehicle groups had similar attitudes
towards the environment as the agencies (1981, pp.
176-179). In view of the superior financial
resources of the producer groups this result may
seem to be surprising, and to indicate that, in
spite of early weaknesses in their history, these
major agencies of government appear to have struck a
balance between the lobbyists in matters concerning
land use.

ZONING BY MUNICIPALITIES

     However, it was at the level of the
municipality, rather than that of the federal
government, that the regulation of land use
generally began in the United States; and one of the
earliest controls was zoning. Some cities and towns
introduced zoning ordinances as early as the 1880s,
but widespread adoption of the practice only
occurred after its introduction in New York in 1916,
a favourable judgement in the case of the Village of
Euclid v. Ambler Realty Co. in the United States
Supreme Court in 1926, and the subsequent Enabling
Act passed by Congress in 1928. Under this law, and
similar Acts passed by state legislatures, the
effective control of land use was transferred from
the states, which had made little or no use of their
powers, to the smallest units of local government -
the municipalities and townships(1) - which were
permitted thereby to limit the types of development
on land within their boundaries. Thus the process
by which control was passed from the highest level
of government in the United States to the lowest was

completed, and the decades since the 1920s can be characterised as being a period during which the states and the federal government have made increasingly-strenuous attempts to control the practices which have grown up following the devolution of these powers, and to curtail the powers themselves: in short, to take the powers back.

The chief instrument of zoning is the town map. A typical map will show that all the land in the municipality will be zoned for some kind of development, even though it may be in agricultural use or under forest. It will also indicate that, while there may be restricted areas in which some noxious activities may be permitted, much of the municipality will often be zoned for residential development, often at low densities. Some land may be excluded from development because it is wetland, and historic districts may be delimited in the built-up area, thus indicating that any redevelopment there may be subject to additional restrictions. An example of such a map for a township on the extreme edge of the Boston metropolitan region in Massachusetts is shown in Figure 6.1, together with a simplified diagram of the chief zones and current extent of development in Figure 6.2. It may be seen that in this case about a third of the township is zoned for single residences, each occupying a one-acre lot, and that almost all the undeveloped land is in this zone. Such a map is approved by the elected representatives of the municipality, or sometimes by the voters directly after public discussion, and changes in it, or 'variances' as they are known, are granted by independent boards.

The purpose of zoning is clearly and narrowly defined in the enabling legislation. It is intended to protect the citizen from some of the physical dangers arising out of uncontrolled building by promoting "the health, safety, convenience, morals and welfare of the inhabitants" (Massachusetts 1982, section 2). This is generally taken to mean the need to restrict the density of building, especially where many structures are of wood, as a safeguard against fire, and to secure adequate daylight. It also allows the segregation of incompatible land uses, the limitation of congestion and the better control of traffic in towns, with the result that life is safer, cleaner and more pleasant than it would otherwise have been for Americans if zoning had not been introduced (Williams 1981, vol. 5, pp. 425-427). However, the early cases of zoning

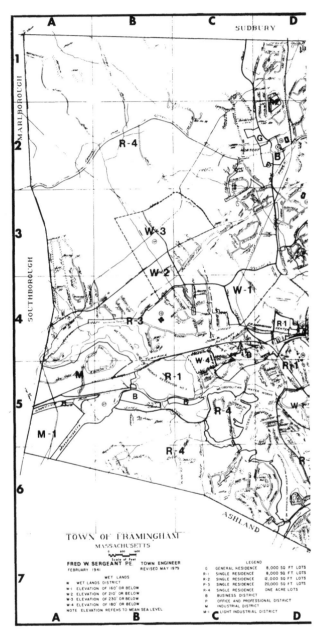

Figure 6.1: Town Map of Framingham,

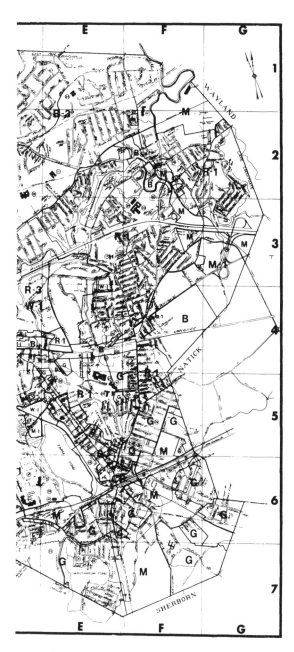

Massachusetts in 1979

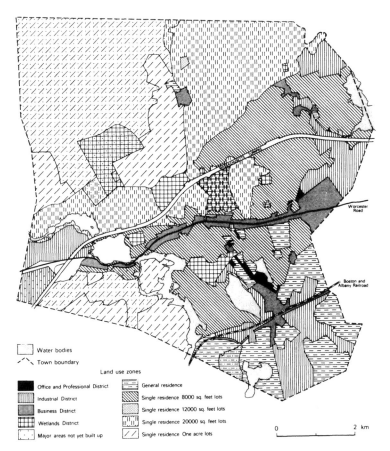

Figure 6.2: Land-Use Zones and the Built-Up Area of
Framingham, Massachusetts in 1979

in San Francisco and New York were designed to
achieve very different ends. The diffusion of
ethnic minorities and industry into more fashionable
areas threatened the property values of the
well-to-do; and, after the passage of the Enabling
Act, the local people who were elected to draw up
and administer zoning regulations, and to grant
variances, rapidly exploited them in ways which
would protect those values.
    Firstly, and most controversially, zoning was
used to exclude certain groups in the population
from municipalities or parts of them. In
particular, blacks were explicitly excluded until

the courts struck down such racialist practices.
The poor were also prevented from moving into
townships by large-lot zoning, that is, by requiring
that all lots be of, say, one acre or more. Such
demands raise the costs of development and
necessitate the construction of detached houses
rather than cheaper apartment blocks. The effect of
this practice was strengthened in some townships by
the imposition of subdivision regulations which
required large floor areas for houses or
unnecessarily high standards of road construction
for new developments, with the result that
low-income groups, including racial minorities, were
excluded. Although the effect of these practices
was often racialist it was possible for townships to
argue that they were in the interests of the
residents because they reduced the calls upon public
services from poor and disadvantaged groups, thus
ensuring that the property tax would be low. This
defence proved to be more resilient in the courts
than outright racial discrimination, but it was also
effectively overturned in several states during the
1970s.
        A second and closely-related practice has
concerned 'ratables'. Township boards and councils
have realised that the property tax could be
minimised by excluding or zoning out land uses which
yield little or no income, such as low and
moderate-income housing in multiple dwellings or
mobile homes, and by attracting activities with high
ratable value but which make little call upon local
services, such as supermarkets, office blocks and
research-and-development institutions.
Consequently, large areas are often zoned for 'good
ratables', and applications from developers for
variances which will allow the introduction of such
land uses are considered favourably. Examination of
the sources of revenue of the various levels of
government in the United States reveals the extent
of the pressure on local authorities to act in this
way. In the country as a whole almost all local
government revenue was raised from property taxes
until the 1970s; and, during a period in which the
range of public services provided by the authorities
has been increasing, and in which inflation has been
rapid, the fiscal problems of those authorities have
been considerable. Indeed, federal and state
governments were obliged to relieve local government
of some of the burden, but even so three-fifths of
the locally-generated revenue of municipalities and
townships at the end of the 1970s, and a third of
all their income, including inter-governmental

transfers of federal and state funds, were from that
source (United States Department of Commerce 1980).
What is more, the proportion remained very much
higher in some areas, and the pursuit of 'good
ratables' in these communities has persisted.

The consequence has been that American society
and land use have become more segregated over space
than ever before. The mobility which has
accompanied mass car ownership - a phenomenon which
occurred in the United States before any of the
other countries in this study - has allowed the
growth of suburbs and the development of a
discontinuous splatter of housing over a wide rural
area around many cities. But the migration to the
suburbs has been selective, for the poor have been
obliged to remain in the cheaper, rented housing,
built largely before the First World War, which now
forms much of the inner parts of the built-up area,
and which has become the ethnic ghetto.
Furthermore, as jobs have migrated from the inner
cities or disappeared through factory closures
unemployment and poverty levels there have risen
further, the possibilities of moving out have
diminished, the services required by the population,
and consequently the level of the property tax, have
risen, and more land has been vacated, in a vicious
downward spiral since the 1950s. This consequence
of land-use regulation has been possible because the
long-established pattern of township boundaries no
longer accords with the functional structure of
society. The jurisdiction of many cities does not
extend into the suburbs, but covers only the densely
built-up areas around the central business district;
and thus the suburban municipalities, which have
been the chief practitioners of both exclusionary
zoning and the search for 'good ratables', have been
largely able to prevent the outward movement of the
poor, the blacks and the hispanic element in the
population, and to preserve their character as
extensive, low-density residential areas.

It is not surprising that this segregation has
produced much controversy. Nor is it unusual that
the activities of the zoning authorities should have
been subject to frequent challenges in the courts,
for, unlike the British, the Japanese and the
peoples of Eastern Europe, Americans have always
made extensive use of the judiciary in the
settlement of disputes over the regulation of land
use. Before 1928 most courts were hostile to zoning
in principle, though some was allowed where it
clearly promoted public health and safety; and even
after the **Enabling Act** of that year the courts

usually upheld requests for variances from the zoning map by developers. Thus, it was not until after the Second World War that courts began to uphold zoning ordinances as a matter of course, or to support the municipalities, but since then the courts have generally upheld refusals to permit variances unless developers could give them good reasons for such variances. It is ironic that in following this more resolute approach they often supported zoning ordinances whose effect, and probably whose intent, was exclusionary (Williams 1981, vol. 1, pp. 117-120). However, many exclusionary practices were struck down during the 1970s; and one of the clearest statements of the judicial view at that time was given in the case of Southern Burlington County N.A.A.C.P. v. Mount Laurel township in the New Jersey Supreme Court. The Court held that it was unlawful for a township to exclude low and moderate-income housing, even if there was no demand within the township for such accommodation, because it was unreasonable to ignore the requirements of the larger region around the township. Conversely, other courts have increasingly taken the view that, where zoning ordinances are based upon properly-contructed and comprehensive plans for the efficient and orderly expansion of public services, which have been approved by the community and from which it is possible to deduce the zoning map, they should be upheld. In several cases during the 1970s, and especially in Golden v. Ramapo in New York State in 1972, and in the Construction Industry Association v. City of Petaluma in California in 1975, courts have supported ordinances whose aim was the control and management of rapid population growth. What is more, they have done so, even though the ordinances failed to acknowledge in full that there was a wider, regional dynamic behind the growth, which could not be dealt with adequately within the existing pattern of local authority boundaries, and despite the fact that they could be exclusionary in effect, on the grounds that they were based upon properly-constructed plans, and were not, as so many ordinances are, essentially arbitrary. Furthermore, ordinances have also been approved recently for no other reason than that they have been passed by local referenda (Callies 1981, p. 293).

The courts have also become involved with the fiscal pressures behind exclusionary zoning. Those in California and New Jersey have held that locally-based school financing, which is dependent upon the property tax to a large extent, is against the Fourteenth Amendment of the United States Constitution because, while this amendment guarantees equal protection to all citizens, a system of locally-financed schools often fails to provide the same opportunities for children in all school districts. The provision of a minimum standard of education requires the sharing of public revenue between rich communities and poor, and therefore the courts in these states have put an end to the fiscal independence of municipalities, and to their dependence upon the property tax as the chief form of revenue, although they have not, by so doing, equalised the levels of property tax between neighbouring townships.

Thus the courts have gone some way to remedy the most glaring perversions of zoning, and there may be important consequences arising out of their decisions for the way in which land is used. However, their influence should not be exaggerated. For instance, it should be noted that, except where matters have been pursued to the United States Supreme Court, decisions have only been binding in the states in which they have been handed down. Furthermore, the courts have not yet been called upon to deal with a variety of other problems associated with zoning which have given rise to a variety of land-use issues. In particular, success in attracting good ratables has led to a sprawl of out-of-town restaurants, shopping malls and low-density housing around towns both large and small. These developments have not only taken over farmland and often created ugly approaches to towns, but have also caused the value of adjacent farmland to rise to such an extent that farmers have not been able to pay the property tax on it, and have been obliged to cease farming or to sell to speculators or developers. The result has been that more land has fallen idle in the urban fringe in some parts of the country than has been developed (Coughlin et al. 1977). However, zoning boards have been very unwilling to establish 'agriculture only' zones, and thereby upset the common expectation on the part of rural landowners that agriculture and forestry are temporary uses only, and that ownership carries with it the right to the proceeds from eventual development. Such action as has been taken to deal with the problem has come not from the courts, but rather from state legislatures, as we shall see in a later section of this chapter. Similarly, the delimitation of historic districts in the centres of the cities has given rise to some consequences of dubious value which have gone largely unchallenged. When the city of Charleston, Virginia, broke new ground in 1931 by establishing an Old City District covering almost all the central business and decaying, port areas of the town it was concerned to preserve their eighteenth-century character by preventing the demolition or uncontrolled alteration of buildings of that age. Height limits were also placed on new buildings to ensure that they would be consonant with the historic townscape. However, as the public has come to appreciate the attractions of the area, it has invested in it, renovated it, raised the value of property in it, priced the original poor inhabitants out of it, and filled it with gift shops, boutiques and up-market restaurants (Ford 1979, pp. 213-216); and many of the cities which have followed Charleston's lead have experienced similar results.

However, it should not be assumed that all parts of the United States are afflicted by the problems, or have experienced the land-use consequences, which are often associated with zoning. Indeed, it has been argued that what is required is not less, but more and better zoning. A survey in 1976 noted that less than ten thousand of the 35,500 municipalities and townships in the country had adopted zoning ordinances, and that in theory any type of development was permitted anywhere within the boundaries of the other communities. The lack of an ordinance is usually the result of an absence of development pressures, and many rural areas with declining populations have felt little need for such controls. But the large scale of some recent developments, and the location of them in areas which previously had experienced little economic growth, have exposed the problems which such a lack of control can create. The expansion of mining in the sparsely-populated states of Colorado, the Dakotas, Montana, Utah and Wyoming has been opposed by many of the communities on the ground that they are required to provide public services for the influx of workers and their families, but in the absence of pre-existing zoning ordinances it has been difficult to prevent such developments occurring (Jackson 1981). Similarly, the growth of population in the smaller towns and unincorporated areas of 'micropolitan' America has gone on largely without planning in the absence of effective local zoning (Chow 1981).

Several other powers are also available to local communities which wish to regulate the way in which land is used, even where a zoning ordinance has not been adopted. For example, their

responsibility for the provision of public services
such as roads and water supplies offers many
opportunities to steer development to some places
rather than others; and frequent scandals over the
supplying of such services to land belonging to
members of local boards or their friends, rather
than to more appropriate sites, are testimony to the
widespread existence of a belief that such
investments do influence land values and,
ultimately, the pattern of development. Secondly,
sub-division regulations and the building code
provide detailed constraints on the design and
layout of developments; sewage-soakaway permits can
be used to reinforce a desired pattern of land use
in unzoned areas; and unwanted developments can be
delayed by negotiations over the granting of the
many different kinds of permits which are required,
and perhaps discouraged altogether. Nevertheless,
the general pattern of development is determined by
the zoning map, and is most likely to be controlled
only where such a map exists.

THE FEDERAL INFLUENCE

    At the opposite extreme of government, the
federal authorities also influence land-use
patterns. However, there is no overall land-use
policy at this level of government, for all attempts
to pass a general land-use law through Congress in
the 1970s failed. Rather, there is a vast array of
Acts of Congress and administrative procedures which
are all designed to affect the use of particular
types of land, or of land-use activities, in limited
ways. Many are not primarily concerned with
questions of land-use planning at all, but with
problems such as the protection of navigation, the
provision of decent housing or the maintenance of
farmers' incomes. On the other hand, much
legislation was passed during a brief period at the
end of the 1960s and early in the 1970s which was
intended to enlarge the federal powers to control
land-use change. The upsurge in legislation at that
time was in response to a widespread acceptance of
the view that zoning had failed to cope with a wide
variety of external costs arising from the use of
land, and especially those which constituted
nuisances not only to the immediate neighbours, but
which affected people and property far beyond the
boundaries of the townships in which they were
situated. However, Congressional scope for action
in this matter is severely limited by the fact that,

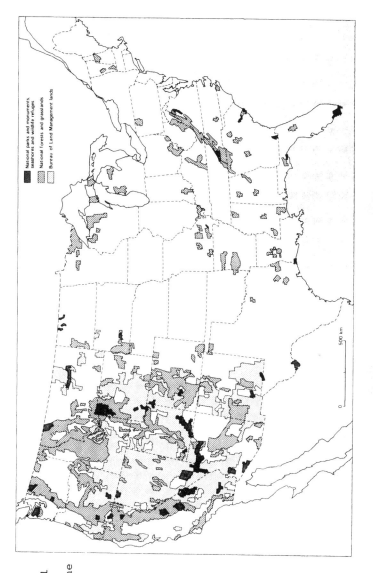

Figure 6.3:

Land in Federal

Ownership in the

United States

under the Constitution, the regulation of land-use
is the responsibility of the states, except in those
areas which have remained in federal ownership. As
a result, policy in Washington is only as effective
as federal persuasion; and persuasion has usually
required the offer of central government funds to
states, localities and individuals who are prepared
to use land in particular ways.

The greatest opportunities for the federal
government to introduce land-use regulation have
been in those areas which have remained in its
ownership (Figure 6.3). We have already noted the
establishment of the National Park System and the
forest reserve, and the withdrawal of land from
homesteading before the Second World War. However,
several other steps have been taken in more recent
years to limit the ways in which federal land can be
used. Since 1960 the Forest Service has been
charged under the Multiple-Use Sustained-Yield Act
with the duty to manage its forests in a way which
will not only ensure continuing timber production,
but also permit their use for grazing, hunting,
recreation, the preservation of wildlife, and
watershed protection. This instruction covers six
percent of the country. Secondly, the Wilderness
Act of 1964 established a National Wilderness
Preservation System, and set in motion a review of
all the so-called 'primitive areas', 'roadless
regions' in national parks and 'wildlife refuges' in
excess of five thousand acres, in order to assess
their significance as wilderness, and to make
recommendations as to whether or not they should be
incorporated into the System. Designation of land
as wilderness results in the restriction of
lumbering and mining operations there. These
procedures were extended to lands under the control
of the Bureau of Land Management by the Land Policy
and Management Act of 1976, and, as a result of
this, a fifth of the country, chiefly in the West
and in Alaska, is now covered by the review, and
many new wilderness areas were designated during the
late 1970s. Thirdly, all developments on federal
land and all other federally-financed developments
are now subject to environmental impact statements
under the National Environmental Protection Act of
1969, before a start can be permitted. In the
statements all environmental resources which are
worthy of being protected must be identified, and
the scope for federal intervention to prevent
development or to steer it away from, say, prime
farmland, sites of ecological or historic interest

and areas liable to inundation by floods under this legislation appears to be great. However, the statements are advisory only, and many pressures can be exerted upon government to ignore them.

Amongst other federal actions which have had the effect of regulating the use of land, irrespective of ownership, some of the most important have been concerned with water. Under the Rivers and Harbors Appropriation Act of 1899 the federal government requires that any construction in or over navigable water, or any excavation or deposit of material in such water, must be approved by the Army Corps of Engineers, and it allows tidal areas to be federally regulated without compensation for any loss of development potential. Moreover, since 1970 the Corps has been required to consider environmental as well as navigational matters in deciding whether or not development shall be allowed. Congress has also prohibited certain kinds of development in a quarter-mile strip of land along each bank of the 392 miles of river which are considered to be of national significance under the Wild and Scenic Rivers Act of 1968, and it ordered a review of a further 4650 miles to assess whether it should also be subject to development restrictions. However, one of the most widespread of the federal government actions with regard to water has not involved any direct control over land use. Since the passage of the Flood Disaster Protection Act of 1973 federal funds for land acquisition and for the construction and repair of public facilities and housing in flood-prone areas have only been available to communities, and to individuals within them, where those communities have imposed controls on further development on land which is liable to inundation by a hundred-year flood. Furthermore, membership of the National Flood Insurance Program, which subsidises flood insurance to those in flood-prone areas, is restricted to communities with such controls. Townships and municipalities are not required to follow any national policy in this matter, but by 1976 over half were already in the process of imposing the approved controls (Soil Conservation Society of America 1977, p. 387). Similarly, the federal Coastal Zone Management Act of 1972 offers to meet up to eighty percent of the expenses of states which draw up and administer coastal management plans in which areas of environmental concern are designated for protection from development. Thirty-five coastal and Great Lake states were eligible for grants under the Act, and, in spite of the requirement that states find

some of the costs themselves, plans from twenty-five
had been approved by 1980.

The federal government has also sought to
influence the location of certain privately-owned
economic activities. For example, it has attempted
to direct industry away from areas in which it could
inflict serious external costs upon the general
public. The Clean Air Act of 1970 and the Water
Pollution Control Act of 1973 make it obligatory for
would-be polluters to obtain permits from the state
government in advance of any development which would
lead to pollution, and thus give states considerable
power to influence the pattern of development.
Similarly the Surface Mining Control and Reclamation
Act of 1977 prohibits opencast workings in the
National Parks, in wildlife refuges and designated
wildernesses, and on alluvial valley floors of
current value to agriculture west of the hundredth
meridian. Moreover, prime cropland throughout the
country is also subject to protection under the Act,
and the states have been made responsible for
ensuring that all land is restored to at least the
level of capability for use which they possessed
before mining began. It has been estimated that
between a quarter and a fifth of all the strippable
coal reserves could be frozen by this legislation,
chiefly in Illinois, Indiana, Kansas, North Dakota
and Missouri, but that the output of coal will not
be affected seriously in the immediate future
(Anderson and Harper 1978, pp. 5-8).

Central government has also tried to tackle the
problem of spillover effects across local authority
boundaries by offering financial assistance under
Section 701 of the Housing Act of 1954 to establish
regional planning councils; and it has linked
assistance to local authorities under about two
hundred different programmes dealing with housing,
economic development, flood control, irrigation,
pollution control, reclamation, transport and water
supply since 1969 to the submission of an A-95
review. This document indicates the extent to which
the developments for which support is being sought
have been coordinated with plans at the regional
scale, that is, at a scale which is larger than that
of the individual municipality, but smaller than
that of the state. Such plans are often drawn up by
counties or other authorities - which are larger
than municipalities, and so are not subject to the
same degree of local pressures over zoning - and it
is the assumption, though not the rule, that
proposed developments which are not in accord with
the plans will not receive federal assistance.

However, all these constraints on the land-use decisions of the municipalities are only as powerful as the size of the grants which are made available. For instance, those for low-income and public housing have always been much less than the value of the tax rebates given to house buyers, and thus the federal government has been supporting the suburban, exclusionary lobby to a greater extent than its opponents.

Indeed, federal tax policy has helped to create a number of the problems associated with the expansion of the urban area. The sprawl of new housing at low densities has been encouraged since the Second World War by allowing both mortgage interest and property taxes to be tax deductible, and by granting larger tax reductions for investment in new housing than for the improvement of existing property. Moreover, federal mortgage guarantees on houses for war veterans have been tied to a requirement that such houses should be detached, and in areas where their value will not decline - that is, in the suburbs. All this has helped to achieve a rate of house building since 1960 which has been faster than that of household formation, and that at a time when that rate was itself increasing; and it has also enabled many people to move to the large retirement settlements which have sprung up in the sunbelt states of Arizona, California and Florida. Moreover, the expansion of the suburbs has been assisted by the way in which the building of roads has been financed. Ninety percent of the cost of the Interstate Highways, and seventy percent of other highway construction, has been paid for by the Federal Highway Trust Fund (Popper 1981), but this fund has been described as a 'money pump', for it is financed by taxes on petrol and tyres, and therefore is likely to grow as the road system is expanded. The result has been an enormous extension of the highway network, an increase in the accessibility of rural areas to the cities, and growth in long-distance commuting to work. Houses beyond the suburbs, and second homes even further afield, have become common; and, in so far as the tax concessions have been of the greatest value to those with the highest incomes, it may be concluded that this action has assisted the spatial segregation of society by income, suburban sprawl and central-city decay. On the other hand, the federal government has not been unaware of these problems. Assistance has been provided since the passing of the <u>Housing Act</u> of 1949 for slum clearance and redevelopment, and for the provision of housing for the elderly and those in congested accommodation in the central cities; and this aid was extended to non-residential property in 1954. Furthermore, assistance under these and some other programmes has been made dependent upon the provision of housing for those of low and moderate income, and since 1974 it has been given only to communities which have drawn up a land-use plan. However, much of this effort merely cleared land, reduced its price, and made it available to office developers, who subsequently made large capital gains out of the increase in property values which followed redevelopment, while at the same time destroying many working-class communities (Scott 1980, pp. 212-217). What is more, the state of the American central city - with its slums, ghettoes, unemployment and poverty - proclaims that federal-government action has not been sufficient to offset the malign effects of zoning upon the use of land.

No account of the attempts by the federal authorities to influence the pattern of land use would be complete without a discussion of agriculture. Although the government is not concerned with the zoning of farmland, and does not discuss with individual farmers the use to which they put it, the problems of agriculture have obliged it to attempt to shape the pattern of land use both directly and through fiscal measures. In the first place, it has long been aware of the desirability of protecting soils. Severe droughts and floods in the early 1930s revealed that the expansion of agriculture into some marginal areas had been unwise. The government responded in 1934 by withdrawing all the remaining land in federal ownership from homesteading until it could be classified in terms of grazing quality and suitability for agricultural settlement. It established the Soil Conservation Service; and, in the Soil Erosion Act, took powers to require states and municipalities to regulate the use of land without compensation where that land was liable to erosion, in order to protect agricultural potential and to prevent the sedimentation and flooding of rivers. More recently, prime farmland was designated as a significant natural resource under the National Environmental Protection Act in 1976, and federal agencies were instructed to avoid developments on it if alternative sites are available. However, few plans have been modified because of their impact upon agricultural land (Berry 1980, p. 167).

Secondly, in contrast and perhaps in opposition

to these attempts to maintain and protect the agricultural potential of the country, the federal authorities have been obliged to struggle throughout the same period with the consequences of the fact that there has been too much land in agriculture. In fact, crop productivity has grown so much faster than the domestic demand for farm products that the problem of crop surpluses has persisted in spite of a substantial decline in the area of cropland since 1953, and a high and growing level of exports. Federal intervention to deal with the problem dates from the Agricultural Adjustment Act of 1933, which introduced price and income support for farmers; and after the Second World War the government's role was extended in the Agriculture Act of 1956 and the Agricultural Trade Development and Assistance Act of 1954. Under this legislation farmers were able to withdraw their land from cultivation in return for government payments, and the government subsidised exports of agricultural products. However, price-support measures and, after 1965, disaster payments to farmers to compensate for crop failures, had the effect of encouraging too much production, some of which was in the wrong place. In addition, the volatile behaviour of demand in overseas markets in the 1970s led first to a rapid increase in the area under crops in the United States, and second to a period of growing surpluses. The Food and Agriculture Act of 1977 attempted to give further support to commodity prices as the surpluses grew, but the Reagan administration found the cost of the cereal and dairy support programmes too high, and reduced the federal subsidy. Much land which had been kept in use by the various support measures was being idled in the early 1980s, and the pressures on farmers were particularly acute in marginal and high-cost areas.

VARIETY AMONG THE STATES

Thus the federal government has built up a wide range of powers which can be used to influence the use of land in the United States; but it has failed to provide an overall framework of law within which other tiers of authority can operate. As a result, state legislatures have reacted to the land problem as they have perceived it, and a wide variety of controls, both of degree and type, have been developed. A compilation of the powers held by individual states in 1975 (Fuguitt and Voss 1979) shows that all of the forty-eight contiguous states had enabling legislation to permit both townships and counties to zone, but in only nine was local planning mandatory, and in only six was local zoning required. Forty had adopted tax measures designed to give relief to owners of agricultural land and open space who were willing to forego the capital gains arising out of subdivision and development for a period, and thus to keep land from being idled in advance of development; thirty-three had taken powers to determine the siting of power stations; and a later survey shows that by 1982 forty-four offered tax incentives to those holding forest land to maintain it in that use (The Council of State Government 1982, pp. 79-91). However, by 1975 only thirteen had drawn up rules governing the designation of areas of critical-environmental or historical concern. States with most controls in the mid-seventies were in New England and on the mid-Atlantic coast, and also included Florida, Minnesota and Montana. Those with fewest were in the deep south, but also included Kansas, Utah and West Virginia. Most of the controls did not deal with land-use control in general, but with particular problems of inter-jurisdictional spillovers, and almost all of them had been put in place since the mid-1960s (Fuguitt and Voss 1979, pp. 86-87).

A few states which have been innovators in land-use control have received a good deal of attention in the literature, and amongst these Oregon has some of the most far-reaching legislation. Problems of suburban sprawl across the prime agricultural lands of the Willamette Valley, recreational developments along the scenic Pacific coastlands, and a rash of subdivision in the arid rangelands of the east of the state led to the passage of legislation in 1969 which required all local authorities to prepare comprehensive plans as a basis for zoning ordinances. This was followed by the Land Use Act of 1973, which laid down that these plans should be in accord with state guidelines, and that they should be submitted for approval to the state's Land Conservation and Development Commission. Amongst the guidelines are requirements that all cities should zone the areas which will be needed for growth up to the year 2000, and that all agricultural and forest land outwith the boundaries of these areas should be zoned for exclusive agricultural use. Unique wildlife habitats, wetlands, wild rivers and lakes are also protected by the guidelines. Results from the 1978 Census of Agriculture suggest that the legislation has been

successful in preserving farming in those areas
which had been subject to the pressures of suburban
growth, in spite of a further rapid increase in
population (Furuseth 1981, pp. 62-69).

Few other states have advanced as far in their
controls, but one which has tackled its growth
problems with vigour has been Vermont. Prior to the
1970s many communities had not bothered to draw up
zoning ordinances, but when the state was suddenly
made accessible to major urban areas by the
construction of the Interstate Highway system it
experienced an upsurge in proposals for the
development of second homes, some of which were for
projects on a very large scale. The state
legislature responded with the <u>Environmental Control
Act</u> of 1970, which gave it power to refuse all
except small developments of second-homes in
townships which did not have zoning ordinances, and
all large developments which were found to be
wanting on aesthetic or environmental grounds or
which would incur unacceptable public expenditure on
the provision of schools or water supplies in other
townships. However, no exact standards were
specified, and there is no inspection or enforcement
system. The <u>Act</u> also empowered the state to draw up
a Land Capability and Development Plan in 1973,
which permitted local communities to plan for and
regulate their rate of growth, but only in cases
where such control would not have a substantial
impact on surrounding townships. Moreover, the
state imposed a capital gains tax on land which was
held for short periods and then sold, except where
this was at the principal residence of the owner,
thus discriminating against speculators and
large-scale, out-of-state developers. As a result
of all these actions many rural communities have
been rescued from the prospect of being swamped by
development, the quantity of development has fallen,
and its quality improved (Healy and Rosenberg 1979,
pp. 73-74). However, some of it has been driven
into neighbouring New Hampshire, which has no
comparable law, and a tax structure which is more
favourable to developers (Myers 1974, p. 24). It
should also be noted that that part of the <u>Act</u> which
would have called into being a Vermont Land Use
Plan, in which permitted locations for development
would have been shown, and which would have affected
the development rights of the residents of the
state, was heavily defeated by the electorate.

Other innovating states have been notable for
pioneering new methods for dealing with specific,
rather than general, problems of growth management.

For example, in 1972 the voters of California reacted to the increasing development of the coastline by establishing seven regional (that is, sub-state) commissions to control almost all types of construction in the coastal belt, in place of controls by local communities which were only too pleased to attract good ratables. Then, the Coastal Act of 1976 specified that developments in coastal waters, and between the coast and the crest of the coastal ranges, should be subject to the overall control of a state coastal commission working through the municipalities. However, there is no provision for the enforcement of the commission's decisions or for it to inspect developments to see whether they comply. Another innovating state is Maryland. Pressures of suburban growth on farmland there led the legislature to experiment with preferential tax systems as early as 1957, and more recently it has taken control of electricity-generating plant siting by the Power Plant Siting Act of 1971. Under the Act the state purchases the sites which it considers to be most suitable on environmental grounds from amongst those identified as being of interest to the electricity-generating companies, and holds them for sale or lease to the companies, whether or not they accord with local zoning ordinances. It also advises the Public Service Commission of the state as to where new plants should be sited. In these ways the state attempts to steer development to its preferred sites. Yet another innovator has been Pennsylvania. The Surface Mining Conservation and Reclamation Act of 1971 there requires state permission for strip mining and the restitution of the general configuration of the land after mining, but it does not override local zoning. Nevertheless it has resulted in the reclamation of tens of thousands of acres of stripped areas.

But these states are not typical. Not only does the mix of problems with regard to land vary widely between states, but the extent to which state governments have taken advantage of the assistance which has been offered by the federal authorities or the example which has been set by other states differs greatly across the country. Thus the nature and handling of the land problem in the United States is much less simple to describe than in any of the other countries which are included in this study, and, for this reason, detailed accounts will be given of the position in three states which have not previously featured prominently in the literature, but which may be more typical of the

country   as   a   whole   than   those   which   have been
discussed above.

MASSACHUSETTS

    The Commonwealth of Massachusetts is one of the
longest-settled parts of the United States by people
of European extraction. Early settlers moved inland
from Cape Cod and up the Connecticut valley, and
many townships were established as early as the
first half of the seventeenth century. At that time
the   coast was fringed with fishing and   port
settlements,   and   at   a   later   date   large
concentrations of manufacturing industry grew up at
water-power   sites   inland.   Many   educational
institutions were also founded,   some of which
acquired high reputations and became the foci not
only of pure research but also of the commercial
applications of the discoveries which were made.
This long and varied history of settlement has
caused   Massachusetts to   be one of the most
densely-populated states in the Union. Almost all
the   land is in private ownership,   and about
two-thirds of it had been cleared of forest, chiefly
for agriculture,   before the middle of the last
century.
    Today   the situation is very different.   The
westward   advance   of   the   frontier opened up
agricultural lands which are far superior to the
cold, stony, wet boulder clays of New England; and
farming declined in Massachusetts after the 1850s.
Similarly, the staple industry of many towns in the
early twentieth century - textiles - has dwindled as
firms have moved south or failed to compete with the
cheaper products of the sunbelt states,   where
heating   costs   are   lower and labour is less
unionised.   Nevertheless, the population of the
state has continued to grow during this century as
new,   science-based industries have expanded (Table
6.1);   many people have moved into the cities,   and
farms have been abandoned.   However,   since the
Second World War there has been another element
within the pattern of migration, namely, the decline
of the central cities. Boston, the largest city in
the   state,   grew rapidly in the nineteenth century,
and it incorporated a few of the adjacent townships
upon whose boundaries its expansion was pressing.
In this century,   however,   it has exhausted the
supply of developable land within its boundaries,
and has found itself hemmed in by cities and
townships which have been unwilling to be swallowed

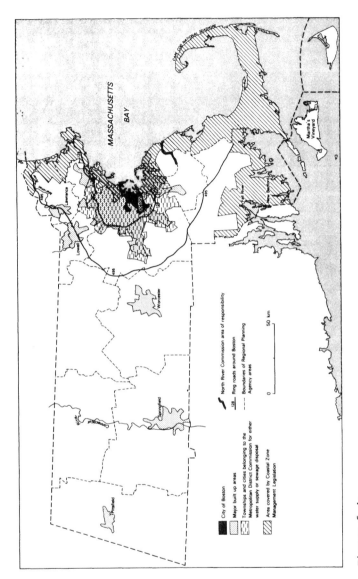

Figure 6.4:

Areas of Some Federal, State and Regional Land-Use Planning Authorities and Controls in Massachusetts in 1981

up.   As a result the city-to-suburb migration has
reduced Boston's population from a peak of 801,000
in 1950 to 563,000 in 1980.   Furthermore,   the
proportion of its population which is black and
living in poor-quality housing has been growing;
property values in the city have been falling in
relation to taxes; many buildings have been
abandoned, and vandalism and the clearance of
property for road-building programmes, which have
since been abandoned, have left much land empty and
unused.   Similar problems have affected the other
major towns in the state, such as Worcester and
Springfield.

    In the same way, jobs have also moved away from
the cities.   The traditional trades of Boston have
declined, and new, high-technology industry in the
form of computer and electronic engineering has
established factories,   research-and-development
laboratories and headquarter offices not in the
city,   but in the suburban townships along the
ring-road, highway 128, and similar activities have
begun to colonise the outer ring-road, highway 495,
as well (Figure 6.4).   The redevelopment of the city
centre during the 1970s has not managed to stop this
trend.   One result of these movements has been that
property taxes, upon which Massachusetts townships
depend to a higher degree than in the United States
as a whole, have soared; and by the mid-1970s a
three-fold difference had developed between the
level of taxes in Boston and those in most of the
surrounding suburbs (Figure 6.5). The significance
of this contrast lies not only in the incentive it
provides for firms to locate outside Boston, but
also in the fact that it is the areas of highest
taxation which are those of lowest average income
(Figure 6.6), while some suburban townships, which
have attracted new development and in which property
taxes are relatively low, enjoy average incomes
which are more than twice those of the city. More
recently the matter has been exacerbated by the
reaction of voters to the high property taxes in
Massachusetts in general. In 1980 they approved
"Proposition 2 1/2", which limits local taxation to
two-and-a-half percent of the capital value of the
land and buildings, and has had the effect of
cutting the level of property tax in many townships,
and thus restraining the level of local government
expenditure, but which has also increased the
pressure on townships to attract 'good ratables'.

    The extent of suburban development and the
impact of low-density styles of building for

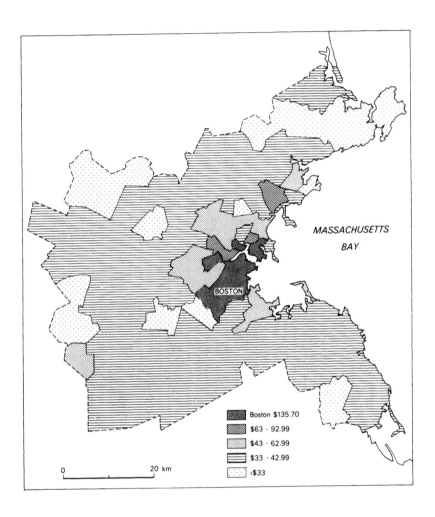

Figure 6.5: Property-Tax Rates in the Metropolitan Area
Planning Commission Region of Massachusetts in 1975

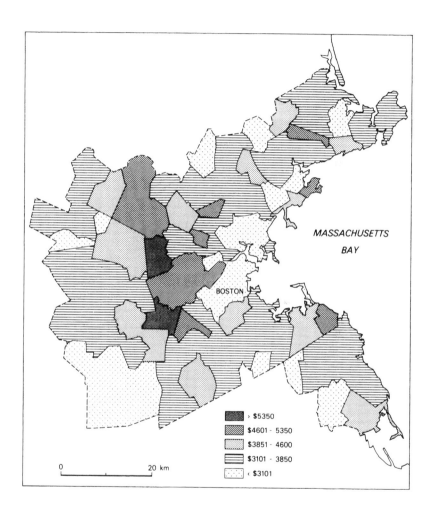

Figure 6.6:   Per-Capita Income in the Metropolitan Area
Planning Commission Region of Massachusetts in 1974
(standardized on 1967 dollars)

housing, industry and out-of-town shopping malls in Massachusetts has been revealed by McConnell. Using aerial photographs from 1952 and 1972 he has calculated that between those dates the residential area almost doubled, and that those of industry and commerce did double, so that developed land increased from seven to fifteen percent of the state despite a rather slow rate of population growth (Table 6.1). Indeed, the urban area grew four times faster than the population. Much of the new development took place in the east of the state on land which had been cleared of forest by 1850, subsequently abandoned by farmers, extensively cut over by the lumber industry in the 1920s, and was approaching the state of mature forest again during the period covered by McConnell's study. However, he noted that the area of mature forest actually increased from twenty to forty-eight percent of the state in spite of the encroachment of development upon it as trees on more-recently-abandoned or cut-over land reached maturity. The area under agriculture, in contrast, continued to fall as land fell idle in the urban fringe and elsewhere, or was developed. Three percent of the state was agricultural land which was lying idle or reverting to forest in 1972, and only six percent was in intensive agricultural use.

Massachusetts has reacted to these changes in a manner which is broadly similar to that of other states. The earliest, and the most important means of influencing the pattern of land use to this day, remains the zoning ordinance of the township. Under the latest state Zoning Act - that of 1975 - all of the 351 municipalities and townships must draw up a zoning map, but the extent to which this represents a planned use of land is limited. In several of the more scantily-populated townships the whole area is zoned for the same types of development, and there is no zoning according to topography, soil quality or scenic attraction. Many do not employ professional staffs of planners, but call on architects or engineers to advise small groups of elected citizens about the zoning map; and only a sixth of the townships employ directors of planning. Moreover, although townships are empowered to draw up Master Plans to guide development, the zoning map need not be related to such plans, or be the product of any stated set of planning policies or goals, so that the designation of zones may be arbitrary.

A variety of agencies have also been created at

a scale between that of the townships and the
state. The most recent of these are the thirteen
regional planning agencies which were established at
various dates between 1959 and 1974. They receive
grants from federal and state agencies, and levy
assessments on townships to pay for small
professional staffs who draw up region-wide plans
for land use, transport, sewerage and water
provision, and for coastal areas of critical
environmental concern. They advise other agencies
at all levels of government, and are the A-95 review
authorities. However, their function is almost
entirely advisory, and membership by the towns and
cities is only voluntary, except for those within
the Metropolitan Planning Area around Boston. On
1 January 1978 eleven of the 250 cities and
townships outwith the Metropolitan Planning Area
were not members of their regional planning
commissions. There are also fourteen counties,
whose boundaries are similar in some cases, but not
in all, to those of the regional planning areas, and
which have some advisory powers over roads.
Thirdly, there have been a variety of regional
authorities in and around Boston. The Metropolitan
Sewage Board was established in 1889, and other
boards were set up soon after for the provision of
parks and water. Initially these served Boston and
about forty of the surrounding cities and townships;
and they were united at a later date to form the
Metropolitan District Commission. This body can
control the pattern of development through its power
to refuse permits to townships and industry which
wish to discharge waste into the metropolitan sewage
system or to take water from the water-supply
network, and to developers who wish to install
electricity, gas or telephone lines. A Metropolitan
Transit Authority was also established in 1947,
which covered fourteen communities. This was
extended to seventy-two in 1964, when it became the
Massachusetts Bay Transportation Authority.
However, this agency depends upon the goodwill of
its constituent members for financial support, and
thus its ability to plan major changes in
accessibility within its area, and to counteract the
outflow to the suburbs, is limited. There is also
the Metropolitan Area Planning Commission, which
covers Boston and a hundred of its suburban
townships, which is the regional-planning and A-95
agency. Thus, there are several different bodies,
all serving somewhat different parts of the Boston
area (Figure 6.4), some of which cover the entire
built-up area and lands beyond, but others of which

do not; and the general picture of land-use control at the regional scale both there and in the state as a whole is that it is much weaker than that at the level of the townships.

In contrast to the regional planning agencies the state is a more powerful body, but yet in comparison with other parts of the United States it is peculiarly weak. The passage of a Home Rule Amendment to the Massachusetts Constitution in 1966 has given townships the authority to exercise any police power not specifically reserved to the state. Moreover, they can veto state plans for transport developments and the designation by the state of disposal sites for toxic wastes. Nevertheless, there are ways in which the state can restrain and influence the pattern of land-use change. In 1959 it forbade large minimum house-area regulations, and thus removed a powerful tool for excluding the poor from townships; and by the Zoning Appeals Act of 1969, which is better known as the "Anti-Snob Zoning Law" it can overrule local refusals of development permits. Under the Act one-and-a-half percent of land must be zoned for low and moderate-income housing, or ten percent of the existing stock must be of this type, or there must be under construction the larger of ten acres or 0.3 percent of the land in the township of such housing, before a township can refuse a permit for housing of that type for other than good technical reasons to a developer who enjoys local backing. Developers are required to demonstrate by way of appeal to a state commission that there is a need for such housing either within the township or a wider area of neighbouring or nearby townships. Many houses have been built after appeals under this legislation, especially in suburban and rural townships, where the stock of such housing is small, and in this Massachusetts has led the way in the United States.

The state has also taken many other powers to control the pattern of land use. It issues permits for all sewer extensions, sewage-disposal facilities, and extensions of water-supply systems of substantial size; and it also requires a separate notification of such proposals under the Massachusetts Environmental Protection Act, which ensures that some attention will be paid to their effect upon natural resources. Through its permitting procedures, which are aimed at the control of pollution, it can affect the location of electricity-generating stations, other fossil-fuel-burning facilities and waste-discharging activities. Particular areas within the state are also subject to controls. Under the _Inland Dredge and Fill Act_ of 1965 (the "Hatch Act") developers must obtain a permit from the state if wetlands are to be affected, and similar legislation (the "Jones Act") covers coastal wetlands. These _Acts_ seek to protect development from flood and storm damage, to ensure that the ground and other water sources are not polluted, and to maintain inshore fisheries. There was also legislation in 1974 establishing a commission to designate critical areas on Martha's Vineyard, and to control development within them in order to preserve the island's "natural, historical, ecological, scientific or cultural values". Similarly, any developments on the banks of the middle and lower reaches of the North River are subject to review by a state commission (Figure 6.4). More generally, the state offers tax incentives to farmers who promise to maintain their land in agricultural use, and a small programme of purchasing development rights from farmers who have been under pressure to release land for development was begun in 1978. Purchases by the state of the rights would preserve the land in agricultural use in perpetuity, but by 1981 less than a quarter of one percent of the state's farmland had been covered because of the very considerable expense involved.

In addition to these controls it should be remembered that federal legislation also influences the pattern of land use. General laws, such as the _Coastal Zone Management_ and _Flood Disaster Protection Acts_ apply to the appropriate parts of any township or state which wishes to become eligible for federal assistance; and the Massachusetts communities which are subject to Coastal Zone controls are shown in Figure 6.4. There is also the Cape Cod National Seashore, which was designated at an earlier date as an area of national significance requiring protection from development.

Together with the rest of the nation Massachusetts underwent a period of increased interest in land-use planning and control in the 1970s, although it was later and less adventurous in this than the pioneering states already mentioned. However, a change of Governor in 1979, and an increase in unemployment about the same time, led to a shift in policy towards efforts to stimulate economic growth wherever and by whatever means possible. As a result the system of land-use controls in the state has not been extended in any major respect for several years, and emphasis has been placed by the government on the need to

streamline and expedite the permitting procedures
already in existence. Under the present system of
controls all parts of the state are subject to some
regulation of land-use change, but the greatest
degree of control exists over new construction along
the coasts and in the Metropolitan District. In
rural areas controls are largely local, though the
financial influences of the Flood Disaster
legislation and the A-95 review process is strong.
However, any major or controversial development,
such as low-income housing, nuclear or other power
stations, toxic-waste disposal sites or new urban
highways, is certain to require a plethora of
permits from township, state and federal agencies,
which offer these bodies a powerful tool for
delaying, discouraging and ultimately refusing
permission. On the other hand, there is no control
over changes in the use of land between agriculture,
forestry and informal recreation; and in many rural
areas house building may still take place in a
dispersed and scattered pattern rather than in an
orderly expansion of the continuously-built-up
area. In short, the pattern of land-use control in
Massachusetts, like the history of its settlement
and economic development, is varied and complicated;
and illustrates the complex palimpsest created by
the actions of the three most important levels of
government in the United States - township, state
and federal.

MINNESOTA

Minnesota is very different. Unlike
Massachusetts it is a very large state, covering an
area which is almost as big as the United Kingdom.
Moreover, it was only settled in the second half of
last century. At that time it was generally
considered to have a harsh climate and poor soils,
and to be isolated from the rest of the United
States; but it appealed to large numbers of German
and Scandinavian immigrants. The chief means of
arrival in the 1850s and early 1860s was by boat up
the Mississippi to the head of navigation at St.
Paul; and later the railways carried settlers to
other parts of the south and to the west of the
state. From the start the economy was based upon
agriculture, and the long-grass prairies of the
south, the west and the Red River valley were more
attractive than the north and east with their thin,
wet soils of boulder clay, extensive swamps and peat
deposits, and heavy forest cover. Most settlement

occurred under the Homesteading legislation, which stamped a regular, but dispersed, pattern of farmsteads on the landscape; much land was drained for agriculture; and the cultivated area expanded until it reached a peak in the 1950s. Large areas in the north and north-east, in contrast, were left in Indian reservations or in federal ownership; and substantial tracts have never been organised for settlement. Timber was exploited early, and the iron-ore deposits there have been extensively mined during the present century. The recent and extensive nature of settlement throughout the state is reflected in the fact that the density of population is only half of that of the United States as a whole, and that, outwith the Twin Cities metropolitan area, there is only one settlement -Duluth - which has a population of more than sixty thousand. This dominance of the Twin Cities developed early. Both grew up under the protection of Fort Snelling, the first and chief point of defence in the state against the Indians; and, while St. Paul became a port and the state capital, Minneapolis developed as the centre of the milling trade, processing the wheat grown by the settlers, with the aid of the power at the St. Anthony Falls on the Mississippi.

However, this state-wide growth of population has not continued since 1950. In the farming areas expanding scales of production and the mechanisation of agriculture have led to substantial falls in the population, and to the decay of public services. Minneapolis has also declined in population - though not in the number of households - during every inter-censal decade since 1950, and the total has fallen from 522,000 in that year to 371,000 in 1980; and in St. Paul there has been decline since 1960. Land in both cities has become vacant as a result of widespread railway and mill closures, and the outward movement of the major firms into the suburbs. However, there is little ghetto development or vandalism, which has been such a problem in Boston, and much new office and shopping space has been constructed in the city centres since the 1960s with the help of federal, state and city funds. Nevertheless, population has risen rapidly in suburban townships around the Twin Cities since the 1950s as the highway network has been extended, and much land has been developed. Suburban shopping malls now rival those of the central business districts in size. There has also been a very considerable and scattered growth of second homes in the forests, and especially on river and lake

Table 6.2: The Most Important Land Use Problems in Minnesota in 1981 by rank (1 is high)

Problem	County Zoning Administrators	Townships	Cities	Regional Development Commissions
1. Scattered residential and commercial development	2	5	7	1
2. Loss of agricultural land	1	1	19	4
3. Incompatible land uses	8	4	6	11
4. Land Speculation	7	2	17	5
5. Orderly annexation	10	3	3	15
6. Malfunctioning on-lot sewage disposal systems	4	7	12	9
7. Solid waste disposal	5	8	14	8
8. Strip commercial development	9	13	4	13
9. Lakeshore development	3	14	24	2
10. Attracting development	19	17	1	7
11. Railroad abandonment	11	23	9	10

12. Development on poor soils	6	12	10	21
13. Water quality	21	25	16	3
14. Extension of city services outside corporate limits	28	10	5	25
15. Wetland destruction	17	21	25	6
16. State and federal wetland acquisition	23	6	--	14
17. Department of Natural Resources state land classification	26	9	--	29
18. Extending urban services within the city	--	--	2	--
19. Traffic circulation	--	--	8	--

Source: Minnesota, *Growth Management Study,* State Planning Agency, Physical Planning Division, St. Paul 1981, pp. B44-B54.

shores,   since the Second World War;   one in twelve
Minnesotan households owns a second home, and two
out of five a boat. Thus the previous pattern of
general population growth throughout the state has
been replaced by decline in the rural areas of the
extreme west and south-west and in the central
cities,   and by increases in a wide area around the
Twin Cities and in the north.

As a result of all these developments about
forty-four percent of the state was in cropland in
1975,   and a further eleven was in pasture and open
rangeland.   Forest   covered   almost   thirty-two
percent,   and all urban development, roads, other
transport facilities and extractive industries had
come to occupy five percent. A state-government
estimate has indicated that by 1990 a further
two-and-a-half percent of the total land area will
undergo a change of use. Two-thirds of this will
occur as a result of the acquisition of land in the
interests of wildlife management, and all the rest -
amounting to less than one percent of the state -
will supply sufficient land for all new urban,
industrial,   mining,   power-production and transport
requirements.   Much of this latter type of demand
will be at the expense of agriculture, especially in
the area around the Twin Cities (Minnesota 1978,
pp. 4-19).

The reactions of Minnesotans - and especially
of the various levels of government in the state -
to this situation vary widely. Whereas the State
Planning Agency does not believe that the changes
which are taking place will prevent agriculture and
forestry from meeting the expected levels of demand
for their products in the period up to 1990
(Minnesota 1978, p. 19), a recent survey by the
Agency,   in which it invited the regional planning
commissions, the counties, cities and townships to
list the chief land-use problems which they were
facing,   revealed both much higher and much lower
levels of concern (Minnesota 1981, pp. B1-B74). On
the one hand, the replies showed that, whereas many
problems were recognised by all the four types of
authority, some issues were perceived to be serious
problems by one, but not by the others (Table 6.2).
Thus,   "scattered   residential   and   commercial
development" and the "loss of agricultural land"
were generally seen to be important, but the cities
gave   much   higher   priority   to "attracting
development" and to extending their services outwith
the city boundaries than did the other groups, while
townships were more worried by land speculation and

by wetland acquisition and land classification by the federal and state governments. Counties gave high priority to problems of sewage and solid waste, and to the related problem of development on soils which are unsuited to soakaway drainage; and the regional development commissions were more concerned with water quality and the destruction of wetland than the others. It should also be noted that most of the problems which were discussed at length in the case of Massachusetts - including the need for an orderly annexation of townships by cities and for the attraction of new development - were also seen to be important in Minnesota. On the other hand, a seventh of the eighty-five county zoning administrators, a third of the 1,851 township boards, and no less than five-sixths of the 855 cities failed to reply, which may indicate that many of these bodies did not consider there to be any important land-use problems within their jurisdictions. Moreover, it should be remembered that, because those who did reply had an interest in planning or other local-government activity, they may have had a predilection for regulations, albeit at the level of their own authorities, and a more-developed view of land-use issues, than the general public.

Any description of the pattern of land-use control in Minnesota should acknowledge the great contrast between the Twin Cities metropolitan area and the predominatly rural parts of the state. Faced with the problem of central-city decline the state legislature inaugurated what has become known as the 'Minnesota experiment', in which it has done more to tackle the problems bequeathed by the map of township boundaries than any other state in the Union. The experiment began with the <u>Metropolitan Council Act</u> of 1967, which established a regional planning agency to cover seven counties (Figure 6.7), and gave it responsibility for public transport, sewerage and park provision. Powers to decide the location of airports and hazardous waste sites were added later; and it was able to insist on the provision of low and moderate-income housing in suburban townships through the A-95 review procedure. Constituent county, city and township authorities were obliged to comply with its plans and decisions about the provision of public services. The powers of the Metropolitan Countil were strengthened further by the <u>Metropolitan Land Use Planning Act</u> of 1976. Under this <u>Act</u> a regional plan was drawn up which indicates the broad areas and the proportion of each local-government unit in

which development will be permitted in the near
future, and all the constituent authorities have
been obliged to produce plans and zoning ordinances
which conform to it. Thus, only the detailed zoning
requirements are now determined by the smaller
authorities. However, the experiment has not been
restricted to matters of direct land-use control,
for an attempt has also been made to deal with the
underlying problems of the central cities. In
particular, the Metropolitan Fiscal Disparities Act
of 1971 compels the rich suburban towns in the seven
counties to share their revenues with the central
cities in order to ensure adequate funds for the
latter, and to reduce the difference in property-tax
levels between the two. Forty percent of the
increase in the assessed value of commercial and
industrial property since 1971 is contributed by
each town to an area-wide pool, which is then
redistributed according to population and fiscal
need. Furthermore, pressure on the property tax was
also reduced by a substantial increase in 1971 in
the proportion of the expenses of the School Boards
which is paid for by the state. Honey and Eriksen
have shown that these changes have not benefitted
all the cities and towns which are fiscally weak,
and that they have hardly affected suburban,
dormitory townships at all (1979, pp. 11-19). Nor
have they prevented the outward movement of industry
and population to the suburbs; but they have reduced
the tax incentive to migrate, and they did give the
central cities extra resources to maintain the
provision of public services, and thus attract
development to their central business districts
during the 1970s.
    Outside the metropolitan area, in contrast,
land-use planning is only weakly developed, and
there is widespread distrust of local regulation,
and suspicion of state and federal interference with
the use of privately-owned land. Although all
counties, cities and townships have the power to
zone, and are supposed to do so on the basis of a
comprehensive plan, a State Planning Agency study of
the extent of land-use control in 1981 revealed that
many areas have no zoning ordinance at all, and that
many which do have no comprehensive plan (Minnesota
1981, pp. B3-B43). Except in Hennepin and Ramsey
Counties primary responsibility for land-use
regulation lies with the counties, but any of the
cities or townships may also adopt their own
ordinances, as long as they are not more permissive
than those at county level. Powers to zone were

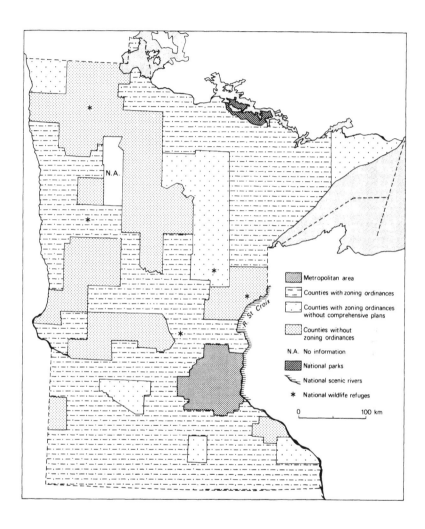

Figure 6.7:   Land-Use Planning by Counties and Federally-
Protected Areas in Minnesota in 1981

only given to the counties in 1959, and the study
shows that only a few had made use of them before
1967. However, forty-seven counties adopted
ordinances between that year and 1973; and
sixty-four are now zoned (Figure 6.7). Most cities
had adopted zoning by the end of the 1970s, but only
a small proportion of townships had done so, largely
since 1975. Regional planning commissions also
exist in most parts of the state, but except in the
metropolitan area they are not mandatory, and are
only advisory.

Thus there is a wide variation in the degree of
regulation of land use across the state at the local
levels of government, but this is not compensated
for at the level of the state by any set of
comprehensive controls. Rather, the state has built
up a huge armoury of legislative weapons which give
it very substantial powers, but only to regulate
selected areas and types of activity. As early as
1973 about 670 distinct sections and subdivisions of
the statutes were found to confer 793 different
powers on departments and agencies of the state
government to control or influence land use (Gilbert
and Gregor 1973, pp. 7-9), and many more have been
added since. Amongst the most important is the
power to acquire land for wildlife management. This
has existed since the 1950s, and about a million
acres are held by the state for this purpose alone,
in addition to its considerable ownership of forests
and parks (Table 6.1). The acquisition of these
lands has given the state extensive control over
both lumbering and mining in them; and, in the case
of wildlife refuges, these activities can be
excluded. Secondly, the **Minnesota Flood Plain
Management Act** of 1969 requires all cities and
counties to adopt an ordinance restricting
development in areas liable to inundation by a
hundred-year flood. Thirdly, the **Shoreland
Development Act** of 1969 obliges the same authorities
to regulate development within three-hundred feet of
rivers and streams, and a thousand feet of lake
shores. Taken together, these controls cover a
large proportion of a state which boasts 17,000
miles of fishing rivers and 10,000 lakes. Moreover,
powers have been taken to identify critical areas,
in which local authorities must make their land-use
controls conform to state plans and regulations; and
by 1979 two areas had been designated and a further
thirty-nine possible ones identified. The state has
powers to control the siting of both
electricity-generating stations and transmission

lines, which it seeks to locate away from prime agricultural land; and all major proposals for development, including subdivisions involving fifty or more lots, should eventually come to its notice through the various laws governing air and water pollution. Lastly, it attempts to discourage the idling of farming by allowing it to be taxed at agricultural-use value.

It is of some interest to note the views of the various planning authorities as to how effective these tools are in the control of land-use change (Minnesota, 1981, pp. B55-B73). The State Planning Agency's survey revealed that, in the experience of the counties, cities and townships, none of them were completely effective, but that zoning, the shoreland ordinance, the sanitary codes governing the disposal of sewage, the floodplain ordinance, subdivision regulations, official maps showing where new streets should be laid out, and the solid waste disposal ordinance indicating where tips could be established were all useful in guiding the pattern of development. On the other hand, environmental impact statements about proposed developments, capital expenditure programmes by government, the Wild and Scenic and Critical Areas legislation and the _Green Areas Act_ to protect farmland from high property taxes were all felt to be relatively ineffective.

This assessment seems to indicate that Minnesota has developed a strong and effective set of controls over land use in areas close to water and in the metropolitan district, but that elsewhere controls are few, limited and weak. There is a widespread feeling that not enough has been done to regulate the mining and lumbering companies in the north of the state, or to grapple with the problem of dispersed development in the six counties which surround the boundary of the metropolitan district, but are still within its commuter zone. Interest in land-use control in Minnesota was strong at both state and local levels in the late 1960s, and for much of the 1970s, but it has declined since then. The State Planning Agency has been trimmed, and the rate of adoption of controls by counties, cities and townships has declined steeply. One of the regional planning commissions has been disbanded.

ARKANSAS

So far the description of land-use planning has concentrated on urban areas and urban influences, but large parts of the United States are rural. Huge tracts, such as much of non-metropolitan Minnesota, lie beyond the direct influence of large cities, with their demands for extensive suburban and ex-urban housing. Nevertheless, as we have seen, where these tracts lie within states which contain very large cities, and where the dominant attitude towards land is that of the urban dweller, restrictions on land-use change may be imposed by the state even where local communities may see little need for zoning or other forms of development control. But Arkansas is not one of those states. Its density of population is low (Table 6.1), and there are few large settlements. The largest, continuous urban area, containing the capital, Little Rock, accounts for about one-eighth of the population of the state, but only contains about 270,000 people.

The settlement of Arkansas began at an earlier date in the nineteenth century than did that of Minnesota, and the state was admitted to the Union in 1836. The early economy was based on agriculture, and especially on cotton plantations on the Mississippi Plain and in the valleys of the Mississippi's chief tributaries. However, the dissected Ouachita and Ozark plateaux remained largely under forest, as they do to this day, with substantial areas in federal ownership. Population growth was halted by the Civil War, in which the state allied itself to the unsuccessful southern cause, but resumed during the period of reconstruction. However, its economy continued to be based upon agricultural activities, and population was widely distributed.

Developments in the post-war period have altered this pattern, but in ways which have been very different from those in either Massachusetts or Minnesota. The decline of agriculture in the 1930s, and especially of cotton planting, was followed by the rapid expansion of the soybean and rice acreages. Soybeans have become the chief field crop, covering more than five million acres, while rice occupies about one million; and cotton has declined from over three million to one-sixth of that area. Farms have been enlarged and mechanised in Arkansas, as in Minnesota, and labour has been shed, but the incentives to produce the new crops have been such that, although the area of farmland has fallen by ten percent in the last fifty years (chiefly as a result of urban growth, reservoir construction and the expansion of commercial timber enterprises), the area of cropland has grown from

5,300,000 acres in 1959 to 8,300,000 in 1979. Much of the wet and wooded batture lands of the Mississippi valley, which discouraged reclamation attempts in the past, have been cleared and drained since 1960 to provide land capable of producing irrigated rice. Thus, while forest remains over half of the state, cropland now covers twenty-three, and grassland nineteen, percent.

The population of the state as a whole fell between 1940 and 1960, but since then it has grown by twenty-eight percent - a faster rate than that of the country as a whole. Much of this increase has been around the larger towns, as manufacturing has been attracted to the state from the north-east, and as employment in government and services has increased. However, almost all of the mountain counties in the north-western half have also experienced more recent population increases, some of which have been caused by the establishment of retirement communities there. In contrast, population has continued to decline on the Mississippi and, to a lesser extent, the Gulf Coast Plains. Urban and other developments have caused some loss of farmland, but in the late 1970s prime land was being converted to other uses at a rate below 0.2 percent per annum, and only about two percent of the land area is built up. Nevertheless, suburban sprawl and city-centre decay have occurred. Although the major cities are not hemmed in by incorporated and independent suburbs whose councils are anxious to encourage shopping-mall and industrial development; although the level of the property tax, and therefore its influence on land-use decisions, has been limited by law for more than a century; and although cities have been able to maintain their populations in the face of the centrifugal movement of inhabitants and employment by extending their boundaries, the downtown decline of Little Rock, for example, and the suburbanisation of both its jobs and population, was accompanied by a trebling of the city's area between 1960 and 1980. Similarly, despite the fact that the recreational demands of the state can be accommodated with ease in the large forested uplands and on the many rivers and reservoirs, there has been some serious river pollution. However, the chief matters of concern are the falling groundwater table under the irrigated farmlands of the eastern part of the state, which may eventually threaten the continued use of land there for intensive cultivation; and the rapid destruction of the batture woodlands, which even modest further increases in the price of the chief crops would accelerate (Shulstad et al. 1980); and these are both problems which stem from the continuing expansion of the agricultural, rather than the urban, area.

In these circumstances it is not surprising that land-use controls in the state are very weak. Zoning was undertaken first in Little Rock in 1937, but planning only began in the mid-1970s. Only about half of the three hundred municipalities with populations of more than five hundred make use of their zoning powers; and none of the counties zone the unincorporated areas. Little Rock expressed a desire in the early 1980s to acquire the power to zone the unincorporated land outwith its boundaries which it will probably annex in the future, because it considered that its existing controls over such land were insufficient to prevent the establishment of non-conforming, mixed and strip developments on them. However, there is little likelihood that the request will be granted by the state or that any effective metropolitan government similar to that in Minnesota will be established. On the other hand, the city can control development outside its boundaries to some extent through the installation of public services, and through subdivision controls. For example, the Suburban Development Plan of 1980 states that the city's power to determine the distribution of sewer and water lines in unincorporated areas will be used to encourage those areas to become part of the city; that areas which require special pumping facilities or which are already approaching complete development will not receive any extension of these services; and that the subdivision controls, which are permitted up to five miles beyond the city, will be used to preserve vegetated areas, and thus reduce runoff and flood risks (Little Rock, 1980, pp. 51-90).

The state government also influences land use in several respects. Firstly, through its Game and Fish Commission it owns about one percent of the land - largely wooded bottomland of the type which has been subject to extensive reclamation for agriculture in recent years - and it assists in the management of about ten percent of the forested lands. In both these areas its chief purpose is to preserve the natural environment from further interference, and to maintain the present vegetation. Secondly, its Septic Tank Law of 1978 limits land sales to sites for which the state's Department of Health has previously granted permits for development, in an endeavour to control the pollution of lakes and rivers. Thirdly, the state is the permitting authority for electricity and gas stations and hazardous-waste disposal sites, and can override local zoning in approving such developments. It offers tax concessions to farmers in areas of rising land values to help them to keep their land in agricultural use; and, like Minnesota, it has the power to draw up restrictions on the development of flood-prone areas where local communities have failed to do so. However, interest in the state, and in the state government, in land-use control is small; the lobbies which favour such controls are relatively weak, and those which prefer freedom for the individual landowner, such as agri-business and lumber companies, are influential. Furthermore, an awareness that on many indicators Arkansas is one of the poorest states in the Union has meant that the economic growth of the period since 1960 has been welcome, that the recession of the early 1980s is something to be resisted, and that any restrictions which inhibit further growth are, at best, suspect. The state administration of the early 1980s, like that in Massachusetts, and the Reagan government in the nation at large, has been more sympathetic to the demands of industry than its predecessors, and it abolished the State Planning Department on taking up office.

CONCLUSION

In an ideal democracy the regulation of land-use should reflect the wishes of the people as a whole; and it may be that, despite the views expressed by Jackson and Strong in the quotations at the beginning of this chapter, the vast majority of the population of the United States is satisfied that the great flowering of controls which has occurred at all levels of government since the Second World War, and especially since the 1960s, has given it a proper degree of control over its land problems. However, this seems unlikely. Even a cursory glance at the serious literature, at local newspapers and television, and at the landscape of America reveals that the regulation of land use is controversial and problem-ridden. While there seems to be broad, but not universal, agreement that a high quality urban environment is desirable, that substantial areas of wilderness should be kept in that state, that local concentrations of air and water pollution should be restricted, that transport

and traffic facilities should be planned in advance
of demand, and that land-use planning has been of
some assistance in these matters in the past, it is
also clear that the detailed control of land use by
government generates many disputes, causes delays to
developers and increases their costs. Furthermore,
it has been used to protect the wealthy at the
expense of the poor, and is widely accused of
encouraging corruption.

The weakness of the present system lies in part
in the multiplicity of authorities which are
involved. Where there is an identity of view
amongst them, and a determination to see a policy
through, land-use regulation can be extremely
effective. On the other hand, even very strong
powers on paper may be of little use if the
authority is opposed by other agencies or levels of
government, if it is short of staff or funds, or if
the general economic and social conditions are
against it at any moment. For instance, strict
federal and state controls over industrial
development are unlikely to be heeded in central
cities or in areas of ecological or scenic value in
stagnant rural economies. Nor are suburban
townships likely to accept federal injunctions to
plan or to provide housing for those of low incomes
if the federal funds which will be jeopardised
through non-compliance are small. The 'sagebrush
revolt' of the late 1970s, in which those western
states in which large areas of land remain in
federal ownership have argued that the federal
government has been too cautious, or alternatively
too insensitive, in allowing development, testifies
to the conflicts which can arise between the tiers
of government (Berry, Parker and Burch 1982, pp.
206-217). In short, there is too much initiative
and flexibility at all levels of authority in the
United States for unanimity of policy to be even
usual, let alone the norm in matters, such as
property and money, which are so close to the hearts
of the population.

Another major criticism of the system of
control is the widespread failure to bring the map
of planning authorities into line with the spatial
structure of the economy. Many problems of
externalities and of exclusionary zoning could be
solved if the primary responsibility for the
provision of roads, sewers and water, the disposal
of hazardous waste, the designation of areas which
are not to be developed, and the determination of
the proportion of houses to be built for people of
low and moderate income, lay with city-region

authorities. This would require the establishment of a whole tier of new bodies similar to Minnesota's Metropolitan Council, with powers to oblige local authorities to conform to the regional plan, and to enforce tax sharing between them. However, there is no sign that the 'Minnesota experiment' will be adopted widely, nor that those regional authorities which have powers will be given any democratic credibility by being transformed into bodies elected directly by the population. Nevertheless, it is important that the authority of such regional councils should be established both on paper and in the public mind, because only then will they be able to take their place effectively amongst the other tiers of government in a hierarchy of units, each of which is of an appropriate scale for the tasks which it has to perform.

In such a chain of command matters of national significance, such as the protection of navigable waterways or of cultural, ecological or scenic sites which are unique within the country, would remain the responsibility of the federal government; and federal protection would override state, regional or township plans. Similarly, the federal authorities would continue to decide whether or not it was necessary to protect the agricultural and forest potential of the country in the light of forecasts about the level of domestic demand and the need for exports, and to determine the consequent level and nature of support required for these industries. Targets could be set for the areas of land which should be kept for these activities in each of the states, and these could be broken down and incorporated into regional and township plans. Alternatively, general financial incentives might be the means of keeping land in certain uses; and land-use targets might be unnecessary. However, it is desirable that the government should definitely adopt a "national referment", as Strong advocates, or that it should not. The present situation, in which there is a general presumption, for instance, against the transfer of prime agricultural land to other uses, but in which there are also a large number of agencies and tiers of government, each with their own priorities, which may have interests in any case, merely gives rise to uncertainty about what land is significant, and engenders wrangling within government as well as between government and developers.

In a clearly-organised hierarchy of regulation the states would be bound to comply with the targets and standards laid down by the federal government,

as they already are in relation to air and water
pollution, and large-scale surface mining; and they
would continue to be responsible for identifying and
protecting sites of state-wide significance.
However, they would be relieved of the need to
supervise the activities of large numbers of
counties, municipalities and townships, at least in
the areas of city regions. That responsibility
could be delegated to the regional councils, which
would be able to enforce regional policies and
plans, as well as national and state standards, more
effectively. Municipalities and townships would
continue to be responsible for zoning within the
broad lines of future land use laid down by the
regions, and for subdivision regulations, but they
would lose their power to determine the timing of
investment in the infrastructure of public
services. Beyond the commuting and recreation
hinterlands of the major cities it might not be
necessary to introduce effective regional
authorities. Indeed, provided that all major
development proposals which are not covered by
special legislation in the manner of surface mining
were notifiable to the state, it would probably be
possible to dispense with much of the patchwork of
regional planning commissions and township zoning
and planning boards in such areas, and to make the
counties the single authorities in land-use and
planning matters. Jackson's comment that there is
"too much planning" would be less justified if the
current ability of the lower levels of government to
resist the plans of the higher, and the present
patchwork of authorities in rural areas, were both
to be reduced.

The establishment of a structured and
hierarchical land-use control system in the form
sketched out above would produce greater uniformity
of practice among different units in the same tier
of government; it would ensure that landowners and
developers were treated in a similar way
irrespective of where they were; it could be
exploited to reduce the number of levels of
government with which any developer has to deal; and
it would diminish the opportunities for sectional
influence and corruption. It would not necessarily
mean that there would be more planning or that the
market would be further weakened as a determinant of
where or when land-use change should occur, but it
would curtail the powers of the townships and
enhance those of regional authorities and of the
federal government. During the 1960s public opinion
began to swing behind those advocating a better

system of land-use regulation and against those arguing for local control, but it is now moving in the opposite direction. Planning agencies are in retreat, and flexibility and economic development, rather than control, in the use of land are the order of the day. The solution to America's problems is being sought in a return, at least in part, to that society's historic belief in the right of the individual to determine his own future, free of all but the more benign and permissive forms of government intervention.

NOTES

1. The terms "municipality", "township", and "local community" are all used in the literature on land-use controls to denote the tiers of government in the United States below the county. They will be used interchangeably in this chapter. The term "local government" usually includes the counties.

Chapter 7

LAND-USE REGULATION IN EASTERN EUROPE

"what _criteria_ are required to make a
'rational' decision? Given agreement on the
criteria, one can then ask: what is the best
decision-making process for selecting among
competing alternatives?   These are the major
problems facing socialist planners today and in
the   future."   (S.L.   Sampson   (1979),
Urbanization - Planned and Unplanned: A Case
Study of Brasov,   Romania, in _The Socialist
City_,   ed. R.A. French and F.E.I. Hamilton,
Wiley, Chichester, p. 521.)

INTRODUCTION

    The countries which have been examined so far
have all been examples of what the World Bank calls
'market economies'. However, the classification of
states in Chapter 1 revealed that amongst the
'next-most-developed' group are almost all the
'centrally-planned economies' of Eastern Europe.
Such a grouping is of obvious interest to a study of
the methods by which the land problem is handled,
not only because of the high priority which is given
in such economies to the communal ownership of the
means of production,   including land,   but also
because of the great emphasis which is placed in
them upon the construction and execution of economic
and other plans by government.
    However, the group is not homogeneous, for it
contains a variety of both form and degree with
respect to the land problem. Although almost all
the countries in Eastern Europe have experienced a
similar,   three-part history of political   and
economic development since the mid-nineteenth
century - beginning with a period of subjugation and

neglect as peripheral elements within the European empires of Austria, Prussia, Russia and Turkey, and continuing through a brief episode of relative independence but economic depression, especially between the two world wars, before falling under the domination of the Soviet Union in the late 1940s - some have followed rather different paths in detail. For example, East Germany was an integral part of the German state and economy until 1945, and for this reason enjoyed a more advanced level of development than other parts of Eastern Europe - a situation which continues to this day (Table 1.1) - while two of the countries which were formerly under Turkish control - Albania and Romania - are still at a lower economic level than the rest of the group. Secondly, it should be remembered that, although Jugoslavia has been deeply influenced by the Soviet model of development, it established a substantial degree of independence from Russian control in the late 1940s, which it has maintained in both political and economic affairs. Thirdly, though most of the countries fall within the 'densely-populated' category in Table 1.2, Bulgaria is relatively land rich. Nor have the histories of the three remaining countries - Czechoslovakia, Hungary and Poland - been identical, but they are characterised by similar levels of development and densities of population.

Thus, there is considerable similarity, but also much variety, among the countries of Eastern Europe which is relevant to the question of land, and therefore this chapter will contain both a general discussion of the supply of, and recent demand for, land among the countries of Eastern Europe, and of the attitudes towards land ownership and the system of land-use control there, and also an examination of the land problem in the largest and most populous of them, and the first in Eastern Europe to draw up a plan for the spatial development of its resources, including land - Poland.

THE SUPPLY OF, AND DEMAND FOR, LAND

The countries of Eastern Europe display a wide variety of resources. For example, Poland covers an area which is substantially larger than that of the United Kingdom, and Jugoslavia and Romania are also extensive states, whereas the others are only about a third of the size of Poland. Secondly, while Hungary and Poland are essentially lowland countries, Czechoslovakia and Jugoslavia are

dominated by mountains and steeply-sloping land.
Thirdly, the pattern of economic development,
especially in the nineteenth century, has left great
contrasts in the density of population and the
degree of urbanisation between different parts of
the area, which have persisted, though to a reduced
degree, to the present day. Thus, East Germany has
both the highest density of population and degree of
urbanisation, followed by Czechoslovakia, while the
lowest are in Bulgaria and Romania. However, within
each country there are both areas of congested urban
development and relatively empty lands. Almost a
fifth of the Hungarian population is concentrated in
the capital, Budapest, together with much of the
country's industry; and the Upper Silesian coalfield
of southern Poland is the largest centre of mining
and manufacturing in Eastern Europe. On the other
hand, densities of population in parts of the
Hungarian steppes remain low in spite of the removal
long ago of the danger of Turkish invasion and the
solution of the problem of water supply there.
Similarly, some of the western and northern lands in
Poland, which were abandoned by their German owners
at the end of the Second World War, are still
relatively scantily populated.

The recent pattern of demands upon these land
resources has also shown great variety (Table 7.1).
While the population of East Germany actually
declined between 1961 and 1980, those of Jugoslavia,
Poland and Romania increased by about a fifth.
Fertility has generally been higher in the
countryside than in the towns, but in all countries
the rapid development of manufacturing industry
since the end of the war has enabled such large
numbers of people to migrate to the cities that
agricultural populations have fallen, especially in
Czechoslovakia, East Germany and Hungary. Pressure
upon urban accommodation has been acute, and the
growth of the built-up area considerable. However,
Ofer (1977, pp. 277-304) has noted that the degree
of urbanisation of the population is low in relation
to the level of economic development throughout
Eastern Europe; that cities have not grown as
rapidly as in market economies in similar phases of
industrialisation; and that a large proportion of
the population has remained in the countryside, even
though many country-dwellers work in the towns.
Thus, the extent of urban growth has been less than
might have occurred; and there may be a pent-up
demand for the transfer of rural land to urban uses
which may have to be met in the future. More
recently, as standards of living have risen in

Table 7.1: Population and Land Use Change in Eastern Europe

	Area (km^2)	Population 1980 (thousands)	Population Change 1961-80 (%)
Bulgaria	11091	8862	11.6
Czechoslovakia	12788	15281	8.9
East Germany	10818	16737	- 2.2
Hungary	9303	10710	7
Jugoslavia	25580	22340	20
Poland	31268	35578	18.7
Romania	23750	22201	19.6

	Agricultural Population (thousands)		Urban Population (thousands)		
	1960	1980	1950	1960	1975
Bulgaria	3823	2947	2001	3005	5067
Czechoslovakia	3495	1571	6354	7886	9580
East Germany	3026	1600	13040	12368	12688
Hungary	3694	1872	3553	4343	5305
Jugoslavia	10324	8348	3269	5242	8327
Poland	11103	10801	9605	14401	19031
Romania	11861	10484	3713	5986	9295

	Change in Agricultural Land Area (%)		Change in Forest Area (%)	
	1961/5-70	1970-80	1961/5-70	1970-80
Bulgaria	5	2.9	2.4	6.2
Czechoslovakia	-0.7	-3.4	0.5	2.9
East Germany	-2.9	-0.3	-0.2	0.2
Hungary	-1.9	-3.6	6.4	9.4
Jugoslavia	-1.3	-2.3	1.6	4.5
Poland	-3.1	-3	8.4	1.6
Romania	1.5	0.2	-1.1	0.3

Sources: FAO Production Yearbook, Rome 1972, pp. 14-15, 19;
1981, pp. 45-74; Hamilton, F.E.I., Urbanization in
Socialist Eastern Europe: The Macro-Environment of
Internal City Structure, in French, R.A. and
Hamilton, F.E.I., The Socialist City, Wiley,
Chichester 1979, p. 168.

Eastern Europe, there has been an increase in the demand for holiday accommodation, especially in the form of chalets in the mountains and forests, or on river, lake or sea shores; and the international holiday industry has expanded along the Black Sea coasts of Bulgaria and Romania, amongst other places. The effects of these changes upon the use of land can be seen in Table 7.1. Between the early 1960s and 1980 the forested areas of all the countries except Romania increased as attempts were made to put back the poorest soils and those liable to erosion under trees. Conversely, the area of agricultural land declined in all the countries except Bulgaria and Romania; and, by the end of the period, Czechoslovakia, Poland and Romania were experiencing increasing difficulties in producing sufficient food to supply their populations, let alone give them the nutritional variety which is normal in developed market economies.

ATTITUDES TO LAND

The communist attitude to land stands in direct contrast to that which prevails in Great Britain, Japan and the United States. In particular, it maintains that land is an asset which belongs to the community, and that as such it should not be in private hands. The reason for this lies in the Ricardian theory that the income of landowners is largely unearned in so far as it arises out of the application of labour and fixed capital to what is otherwise the free gift of nature, and that in consequence private landowners are able to appropriate for themselves any 'surplus' which is made available by production after the payment of some minimum sum which is necessary to ensure the supply of the other factors of production. The charging of what the market will bear for the use of any piece of land - and the allocation of sites to uses by such a method - is therefore an unjustified acquisition by a few, fortunate landowners of the value which society as a whole has endowed the land with, and an expropriation of public wealth for private gain. It follows from such a view that, wherever possible, land should be placed in communal ownership, and that otherwise strict limits should be put upon the size of private holdings.

It also follows that the price of land in socialist societies should reflect not the scarcity value of any site but merely the cost of the fixed

capital, in the form of drains, fertilisers or
public services, which has been invested in it in
order to make it more useful. However, this does
not eliminate the problem of unearned surpluses, or
'rents', for the inherent differences in the quality
of land ensure that, unless the prices of the other
factors of production or of the products vary among
producers, those enjoying the use of better sites
will continue to earn more than those who are
obliged to put up with inferior ones, whether this
be for the growing of crops, the manufacturing of
goods or the supplying of services. Because these
differences exist users will compete for the best
sites, even in a socialist economy, but there will
be little incentive to sell sites to others whose
use might yield a higher rent if the price received
is significantly less than a full capitalisation of
the value of the rent, and none if the land is
free. Indeed, low or negligible land prices are
likely to make for great rigidity in the land-use
pattern, especially where users are sufficiently
powerful to resist administrative demands for
land-use change, and for the extensive and wasteful
use of land in general. Notwithstanding these
problems state enterprises in Poland were not
required to pay for the use of nationalised land
between 1961 and 1969; and, when rents were
introduced in 1969 in order to enforce a more
rational use of land and to provide a means of
ensuring that land-use changes were in accordance
with plans, they were very low. Similarly in
Czechoslovakia land is normally exchanged without
payment between socialist organisations; and it is
common practice in Eastern European countries to
levy much lower rents for publicly-owned housing
than would exist in a free market, irrespective of
the location of that housing in relation to places
of employment or services.

The prices which are paid by the state for
privately-owned land are also low. Indeed, in the
early years of communist control much land was
either expropriated or its owners were forced to
give it, in effect, to agricultural collectives. As
a result, most of the land in Eastern Europe, with
the exception of Poland and Jugoslavia, is in one
form of public ownership or another, although some
remains in private hands. Where privately-owned
sites are required in urban areas in Poland for such
public uses as redevelopment or road widening,
owners are compensated at a rate of between five and
ten percent of the estimated cost of building a
five-bedroom house of not more than 110 square

metres floorspace on the site, or proportionately less if the site acquired is smaller than a standard building plot. The exact percentage which is paid is determined by the cost of building, and varies from the minimum in small towns and suburbs to the maximum in cities and, especially, city centres. In the same way, land which is taken over by the state by pre-emption attracts a fixed price which is only slightly more than that paid for sites which are compulsorily purchased. Some indication of the lowness of these prices is given, firstly, by the very restricted range of values as between plots in villages and those in central cities, which is much less than the variation in prices between comparable sites in market economies, and secondly by the fact that prices of land on the free market in Poland for both urban and rural sites are considerably higher than those paid by the state (Ganguly 1976). Similar discrepancies exist in Jugoslavia between the official and actual prices of land in urban areas (Harrison 1983, pp. 178-181).

However, there has been a change in attitude during the 1970s to the question of land prices. For example, under the **Agricultural Land Protection Act** of 1976 in Czechoslovakia payments must be made for land which is withdrawn from agriculture, and these must compensate the farming organisation in full for any economic loss which it may suffer, and also the loss which the nation as a whole is thought to bear as a result of any reduction in the area of agricultural land. In effect, the payment must be adequate to enable the organisation to maintain the same output and economic results as before the transfer of some of its land (Fabry 1979, p. 87); and the developer must meet the cost of the intensification of use of the remaining land, even if the overall efficiency of the economy would be improved by a substitution of other forms of production for those of agriculture. Similar systems of charges for the transfer of land out of agriculture were also established in several of the other countries of Eastern Europe during the 1970s as concern about the loss of such land grew (Gaidamara 1981, pp. 86-92).

This approach to land pricing appears to avoid some of the most important restrictions on land-use regulation in market economies. Low levels of compensation for the acquisition of private land does away with the problem of recapturing the unearned betterment or development value of land which arises out of investment by public authorities in infrastructure and of the demands of society at large for particular sites. Moreover, compensation for the loss of development rights arising out of zoning does not arise in economies in which almost all major urban uses, including an increasing proportion of housing, are in the hands of the state. However, it removes the bid-rent mechanism, and thus deprives society of a simple indicator as to how land should be allocated among competing uses. Bater has reported (1977, pp. 201-202) that by the mid-1970s a consensus had emerged among Soviet writers that the absence of a fully-developed set of land prices works against rational land-use allocation, and that some surrogate is necessary for the market, based upon the costs of construction on different sites, and upon the accessibility of those sites within cities, the level of their services and the quality of their housing; and similar discussions have occurred subsequently in Eastern European countries (Eberhardt 1980, pp. 527-529).

Thus, an attempt has been made of late to reintroduce market mechanisms into land-use decisions. On the other hand, it has been widely realised that such mechanisms are subject to serious failures, and all the communist governments of Eastern Europe have seen clear justification for substantial interventions in the land market, in addition to those over ownership and land pricing, in order to secure, for example, the segregation of incompatable uses, control over the supply of housing and urban services, and a minimum provision of green and open space. Marx and Engels (1959, p. 28) and Lenin (1961, p. 67) reacted to the problems of the great urban settlements which grew up in the early nineteenth century (and whose counterparts appeared in Eastern Europe later in that century) by suggesting that they should be dispersed, and that urban styles of work and life should be spread throughout the countryside; and early models of socialist cities included parks and sporting facilities. More recently, large areas of attractive landscape around cities have been recognised in plans as providing an important recreational facility (Pallot and Shaw 1981, pp. 203-204, 265-266); and there has been an increasing segregation of land uses in the cities of Eastern Europe since the end of the Second World War, and an evening-out of the intensity of land use, as the density of buildings in congested areas has been reduced and as new housing has been built at high densities in apartment blocks in the suburbs. However, this has rarely been accompanied by the careful location of new industry with a view to avoiding atmospheric pollution in the towns; and the large cities have continued to grow in spite of attempts to restrict migration to them, while small, rural towns have proved to be unattractive to young people.

Communist governments have also accepted from an early date the desirability of protecting natural resources in the long term. For example, national parks and nature reserves have been designated in all the countries of Eastern Europe, though on a much smaller scale than in Britain; and, more recently, attention has been paid to diverting development away from prime agricultural land to other sites (Gaidamara 1981, p. 86). The anxiety over farmland arose at the time of growing world-wide concern over the environment in the late 1960s, and in particular in the early 1970s, when poor harvests in the USSR and the failure of the Polish government to raise food prices led to an increasing dependence of Eastern Europe upon imports of foodstuffs from North America. However, the attitude of the governments of Eastern Europe to the environment has been ambivalent, even during the renewed food shortages of the late 1970s and early 1980s. While atmospheric pollution from Upper Silesia is allowed to kill large numbers of trees in southern Poland, to increase the acidity of rainfall and thus the leaching of soil, and to undo the work of restoring the old city of Krakow, and while Soviet industry continues to pollute Lake Baikal, it would appear that governments place greater emphasis upon production than conservation.

It is also apparent that, in spite of the dominance of government in Eastern Europe, there is another, very powerful attitude towards land among the general population. Some land remains in private ownership in both rural and urban areas, and there is no popular demand for any further extension of public ownership or of the cooperative pooling of farmland. Indeed, the traditional antagonism of peasants to attempts to reduce their control over their holdings is as strong as ever, and one of the primary demands of Rural Solidarity in Poland in 1981 was an acceleration and improvement in the working of the free market in agricultural land, and a reduction in the delays and obstructions imposed upon land transactions by the authorities. Moreover, the authorities in Russia and other countries in which almost all agricultural land has been collectivised or taken into state farms have been obliged to acknowledge that the effort expended by peasants on their remaining private plots is much greater than that upon the collectivised land, and that the output of these plots is essential to the maintenance of adequate supplies of several types of food (Nove 1977, pp. 122-126). Nevertheless, the proportion of the land in socialised uses in Eastern Europe continues to increase as governments take advantage of land sales and compulsory purchase to take over privately-owned sites.

Thus the general attitude of communist governments to land is that it should be used responsibly within a system of communal ownership. By this is meant, first and foremost, the most productive allocation which can be achieved so that the output of goods shall be as large and as efficient as possible. But governments also see it to be their responsibility to ensure that the population has a uniform supply of housing (Ostrowski 1965, p. 45) in a communal environment, and this has led to the widespread appearance of small apartments in large blocks (Reiner and Wilson 1979, p. 65), which is the antithesis of the suburban sprawl of the United States. They have also paid much attention to the need to provide the cultural, environmental and recreational facilities which are considered to be essential to the creation of the all-round, socialist individual (Hamilton and Burnett 1979, pp. 263-270). Meanwhile, individuals adopt attitudes of a more local and self-centred character, seeking to maximise their own, and their family's, returns from the use of land.

THE SYSTEM OF CONTROL OVER LAND USE

It might be supposed that, in an economy in which a very large proportion of the land is in public or cooperative ownership, controls over its use would be less necessary than in any of the other countries which have been examined so far. However, this would not be correct. In the first place, some land remains in private hands in all Eastern European countries in both rural and urban areas, and in Jugoslavia and Poland this includes over seventy percent of all farmland. Secondly, the remaining land, though nominally in national or public ownership, is in fact controlled by a wide variety of state, local and cooperative organisations, each of which has its own interests and policies, not all of which might be in accord with, say, such national aims as the protection of prime agricultural land from development, or the separation of conflicting land uses in built-up areas. However, the centralised, command nature of the economic planning system in Eastern Europe should make the execution of such policies easier. Furthermore, the fact that there are relatively few organisations with which the planning authorities have to deal reduces the likely number of cases in which they will be called on to exercise restraints. Large numbers of competitors for sites in market economies, such as private development companies, banks, insurance offices and professional and personal services, are almost non-existent in socialist societies, and so the pressures which they exert upon land-use patterns are largely absent.

Unlike the situation in the United States, control within the economies of Eastern Europe is simple and hierarchical. As all power flows from and through the communist party, all decisions about the economy, including those on land use, can be taken at the appropriate level in the party machine and transmitted through it. Economic planning is usually undertaken by what Zielinski (1968) has described as the administrative-iterative method. In this, the Central Planning Commission, which is a very senior body of production ministers and other party planners, draws up targets for production during the next five years. After approval by the central committee of the communist party these are sent down to individual ministries and local authorities, who in turn break them down and send them on to their own constituent units. Comments on the targets are subsequently passed back up the chain of command, and the Planning Commission then issues a modified set of targets in the form of a five-year plan, which has the force of law. Similar procedures are also used for the construction of shorter and longer-term plans. Most of the targets which are included in the plans have only indirect significance for land use, for they are concerned with levels of output or the provision of services, rather than with the spatial pattern of resource use. However, others, such as the area to be afforested or mined by open-cast methods, may be directly specified, although the exact location of the land involved may not be given. Moreover, detailed location decisions are made at the highest levels about all major new factories, transport facilities or reservoirs. Figure 7.1 indicates how, in the case of Poland, the local authority's role is restricted to the supply of detailed information to potential developers about the availability of sites within its area. All other decisions are made by the Planning Commission; and only in more minor cases is a decision about location made by individual industrial ministries, or by their branches, or by local authorities. Thus the detailed planning of land use is undertaken to a large extent at an early and elevated level in the chain of command, rather than by local bodies.

Stage of Planning

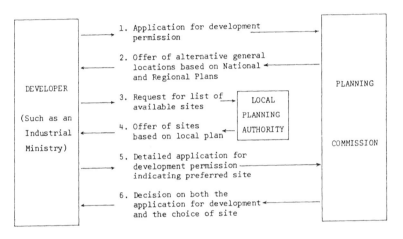

Figure 7.1: The System of Site Selection for Major Developments in Poland

However, the planning of land use does not always sit happily in such a structure of control. For instance, changes in the cities of Eastern Europe are supposed to conform to a local plan, which is designed to provide for the efficient movement of people between residence and workplace, and to ensure the provision of adequate green and open space for recreational and sporting activities, and also for other public services. For example, Pallot and Shaw have indicated (1981, pp. 246-253) that in the USSR the city soviets have great power to control the allocation of land between both individuals and organisations, and can halt construction which violates plans, at least in theory. But they also point out, as does Gilbert-Gebert (1981, pp. 5-6) in the case of

Poland and Kansky (1976, p. 140) in the case of Czechoslovakia, that national government not only determines the policies within which local government acts, but also must approve all major investments in housing and other local services by city authorities. Moreover, they note that the industrial ministries in the USSR are frequently criticised for failing to abide by agreed patterns of development, and in so doing upsetting the land-use planning process at local level; and they quote the view of M.N. Mezhevich that, in consequence, the Soviet city develops as though it were unplanned.

Whether or not that is correct in the case of the Soviet Union, it is certainly too sweeping a description of the situation in Eastern Europe. Notwithstanding the division of economic from spatial planning for much of the period since the Second World War, and the preference which has been given to the economic ministries in disputes with the spatial planners (Jachniak-Ganguly 1978, pp. 117-118), much of the post-war urban growth in Eastern Europe has possessed a good deal of order in at least five respects which are relevant to the land problem. Firstly, most dwellings have been provided in relatively dense, high-rise developments of apartment blocks, in an effort to provide a communal setting and equality in housing space amongst the urban population; and this has prevented low-density suburban sprawl. Secondly, many of these blocks have been provided within planned neighbourhood units, which has allowed the segregation of nonconforming land uses to an extent which was uncharacteristic of the Eastern European city in the pre-war period (Dawson 1971, pp. 109-111). In particular, large-scale industry has been excluded, and each unit includes some local services, and educational and recreational facilities. Thirdly, building controls, including density restrictions, have been widely adopted in order to ensure a full, but not congested, use of land; and fourthly, several cities and conurbations have been surrounded by green belts from which industry and housing is largely, but by no means entirely, excluded. Lastly, considerable attention has been given to the protection of historic towns, especially in Poland. The immense damage during the Second World War there produced a remarkable reaction throughout the nation, as a result of which the ancient centre of Warsaw has been painstakingly reconstructed in its pre-war form, and much of the surrounding central business district of the city

rebuilt in similar styles. The nationalisation of
much of the land in the city at the end of the war
assisted the large-scale planning which this
redevelopment required. Moreover, special controls
over changes in land use and redevelopment have been
applied to the centres of Krakow, Gdansk and Torun,
as well as to many individual buildings and
gardens. In Czechoslovakia there has also been
effective control over changes to listed buildings
since 1958, and thirty-eight town centres have been
recognised as Historic Town Reservations, while in
Hungary thirteen enjoy this status. Proposals for
change to buildings within these areas must be
approved by the national commissions for historic
and public monuments (Dobby 1978, pp. 90-1).
     Nor has there been an absence of land-use
regulation in the rural areas of Eastern Europe. In
Czechoslovakia, orchards, vineyards, hop gardens and
vegetable fields are 'specially protected'; and all
proposals to withdraw land from agriculture by both
public bodies and private individuals must be
approved by both the agricultural and the local
authorities. Only applications concerning land of
lower quality, or land which lies within areas zoned
for holiday cottages, or areas of fragmented
farmland within built-up areas or villages, are
likely to be approved (Fabry 1979, pp. 84-86).
Similar policies were first adopted in Bulgaria in
1967, though they were not effectively enforced
until the 1970s, when they were gathered together
into the <u>Arable Land and Pasture Preservation Act</u>;
and the protection of farm and forest land was
enforced in Poland in the 1970s, thus excluding
development from better soils, confining rural
building within villages, and restricting chalets
and second-homes to limited areas. In addition,
industry and mining have not only been required to
pay for the land which they use, but also to restore
any which may be left derelict in all the countries
of Eastern Europe since the 1970s.

POLAND

     Nevertheless, serious problems have arisen in
Poland in relation to food supply in recent decades
which throw grave doubt upon the system of economic
and land-use planning there. Indeed, since the
establishment of a communist government in the late
1940s there have been several periods of food
rationing; and periodic attempts to raise the
official price levels for food have led to the fall

of the governments of Gomulka in 1970 and Gierek in 1980, and to disturbances at other times. Yet all of this should be surprising for the land resources of the country are substantial, and the pressures upon them in terms of population growth or urban expansion have not been nearly as great, say, as in Japan. It should also be noted that, in spite of the problems which have occurred, it was not until the mid-1960s that any great effort was made to protect the better agricultural land from development, and to subject it to the same types of control as had been applied earlier to areas of scenic and scientific interest in the National Parks.

The land resources of the country may be divided roughly into three. Although most of Poland is low lying, only the southern lowlands of the Vistula-San basin, together with the Lublin plateau to the east, parts of lower Silesia and the Poznan area offer soils of high quality for agriculture. Elsewhere the fluvioglacial sands and gravels and the morainic deposits offer alternately thin, acidic and heavy, stony soils, some of which have remained under forest. Thirdly, there are the steeply-sloping, wooded Carpathian and Sudeten mountains in the south. Only 6.9 percent of the country falls into the category of prime agricultural land, whilst about thirty percent is either forested or considered to be only suitable for that use (Glowny Urzad Statystyczny 1980a, pp. 3, 67), but there is still more farmland available per capita than in either Britain or Japan. In addition, there are substantial mineral deposits below the surface which determine the location of much of the industrial and urban development. Of these, the largest are in the Upper Silesian coalfield, but others include the brown coal and lignite deposits on the East German border in the south-west of the country and at Belchatow and Konin, the metalliferous ores in lower Silesia, sulphur at Tarnobrzeg, and the recently-discovered Lublin coalfield.

The present patterns of land ownership and use reflect the events of the nineteenth and early-twentieth centuries. After the partitions of Poland little economic development occurred in the areas belonging to the Austro-Hungarian Empire, and population pressure increased cruelly in the south-east of the country. The limits of cultivation were pushed into all but the poorest and steepest land, the size of holdings fell, and there was much rural distress and emigration. Indeed, so

great was the pressure on land there, in the absence
of any large-scale industrial and urban development,
that there was little left in the form of large
estates which could be given to the landless
peasants and small farms in either the inter-war or
post-war land reforms. Peasant holdings remain the
smallest and most fragmented to this day in any part
of Poland; and the attempts at collectivisation made
least progress in this area. Similar circumstances
existed in the eastern part of the country, which
lay within the Russian Empire until 1914, although
the pressure on land was not so acute, and the
degree of industrial development, especially in
Lodz, Warsaw and on the eastern edge of the Upper
Silesian coalfield, was greater than in the
south-east. The fortunes of the rest of the
country, however, were very different. As parts of
Prussia they were subject to a higher level of
economic development in the nineteenth century.
Agriculture became a relatively large-scale,
commercial activity; industry was established not
only in Upper Silesia and in the major cities of
Gdansk, Poznan and Wroclaw, but also widely among
the relatively dense network of smaller towns; and
the whole economy was bound together by a
well-developed railway system. Much farmland was in
the form of large estates, many of which were owned
by Germans; and, after the Second World War, many of
these were confiscated by the Polish government to
form State Farms, while other land was given to
Polish refugees from western Russia and to landless
migrants from south-eastern Poland.

Economic policy after the Second World War
concentrated upon the need to remove the surplus
population from the land, and to transform it into
an urban proletariat. To a large extent this was
achieved by expanding the pre-existing industrial
centres, most of which were in areas of only medium
or little agricultural significance. In particular,
the growth of industry and population in Upper
Silesia was undertaken very largely on sandy soils
of low fertility, and the expansion of housing was
restricted by the high-density, apartment-block
style of development. However, some new building
did impinge severely upon the better agricultural
areas, and especially the Lenin Steel Mill at Nowa
Huta, to the east of Krakow. The mill and the new
town, which was erected to house thousands of
workers, covered some of the very best loess soils
on the terraces of the Vistula, while the many
thousands of workers who migrated from south-east
Poland to work in that and other industries in

Krakow, but for whom no accommodation was available, were housed in the villages around the city, and increased the pressure on land there. Partly as a consequence of these developments the proportion of the country in agricultural use fell from 65.6 percent in 1946 to 60.7 percent in 1979, and is expected to decline to 58 percent by 1990. However, not all of this was the result of urban and industrial expansion. One of the most important reasons for this change has been the extension of forestry onto sandy and derelict land, which has led to an increase in the proportion of the country under woodland from 24.5 percent in 1960 to 27.8 percent in 1979, and is expected to raise the proportion to 30.7 percent by 1990 (Jastrzebski 1976, p. 25).

It was also government policy in the early 1950s to extend socialist forms of land ownership as far as possible, and especially in agriculture. However, it proved to be impossible to collectivise private farming in the Russian manner. When the attempt was abandoned in 1956 only a handful of cooperative farms survived, and less than a fifth of the farmland was in State Farms. Since then the proportion of communally-owned land has risen slowly, but it had only reached twenty-five percent by 1970. In contrast, much forest and urban land was nationalised in 1946, as the state took over the ownership of industry and public transport, so that by 1970 eighty-three percent of the forests and seventy-three percent of the built-up, mining and transport land (Grocholska 1972, p. 17), were in public ownership, together with all development rights.

However, land was not seen to be a major constraint upon economic development for most of the post-war period by the Polish authorities. Nor were the spillover effect of development perceived to be serious until the late 1960s. Indeed, it was confidently asserted on many occasions that socialist planning would not only avoid, but would do away with, the legacy of unplanned industrial and urban development of the capitalist era of Poland's history (Prezydium Rady Narodowej Miasta Stolecznego Warszawy 1965, pp. 10-11, 65-98, 129-163). Nevertheless, the effects of the great emphasis which had been placed upon economic growth began to become obvious in the late 1960s, when it was realised that about eight percent of the agricultural area and three percent of the forests were suffering from atmospheric pollution, and that by 1990 about thirty-five percent of the farmland

would be affected (Jastrzebski 1976, p. 26).
Demands were made at the Sixth Congress of the
Polish United Workers Party (that is, the Polish
Communist Party) in 1971 for greater protection and
rationality in the use of the natural environment,
and spatial planning has been accorded greater
importance since then.

However, any account of land-use regulation in
Poland should begin by acknowledging that the
country has had a longer tradition of spatial and
urban planning than most in Eastern Europe, even if
that tradition has been intermittent, and its early
effects meagre. The first, modern, city planning
occurred under the Building Act of 1928 - though
little had been done to put Chmielewski and Syrkus's
1936 plan for Warsaw into effect before the outbreak
of the Second World War - and after the war great
effort was expended upon the plans for the
reconstruction of the city. Similarly, some attempt
was made to plan the location of strategic
industries in the 1930s, and to establish a Central
Industrial Region away from the frontiers; and some
parts of the country were the subjects of spatial
plans during the 1950s. Nevertheless, there was no
nation-wide, long-term system of physical planning
until the Spatial Planning Act of 1961 called for a
three-fold hierarchy of plans at the national,
regional (voivodship) and district (powiat) levels,
and gave these plans a clearly-defined, but
subordinate, relationship to those of the economy.
Under that Act spatial plans were drawn up for the
country as a whole, for each of the twenty-two
regions, for most of the three hundred districts,
and for more than eight hundred towns within the
districts, for a 'perspective' period of fifteen to
twenty years. The purpose of the plans was to
identify the patterns of infrastructure and
settlement which would be necessary if the economic
and social goals which had been laid down by the
central government were to be achieved; and, to this
end, detailed plans of land use were constructed at
the scales of 1:5,000 or 1:10,000 for each district,
while regional authorities were given the power to
refuse permission for the siting of individual
projects where they did not accord with the land-use
aspects of the plan.

This system was modified in 1975 at the time of
the reorganisation of local government, though the
purpose of spatial planning remained the same. A
hierarchy of authorities was established at that
time which is composed of central government,
forty-nine regions (voivodships) and about two

Figure 7.2: Macro-Regional and Regional (Voivodship) Planning Authorities in Poland since 1975. (The name of the chief city in each macro-region is italicised.)

thousand communes (gminy). The new regions are
considerably smaller than those which they replaced,
and they were intended to be city-regions of fairly
homogeneous character, specialising in either
industry, agriculture or tourism, for it was thought
that this would make them suitable to be planning
units. However, because of the decrease in their
size, it was felt desirable to group them into eight
macro-regions (Figure 7.2) in order to facilitate
the coordination of the supply of the chief
infrastructural services. The communes were also
intended to become planning authorities, but only
those with populations of 20,000 or more were
equipped with planning staffs in the first
instance.

The hierarchy of authorities produces a
hierarchy of plans. The most important and senior of
these is the National Spatial Development Plan,
which lays down the broad pattern of urban areas and
transport links between them for the next fifteen
years; while the macro-regional, regional and
commune plans, at scales varying from 1:200,000 down
to 1:5,000, elaborate the details of the proposed
pattern of economic and social development, and the
land-use consequences of them. More particularly,
regions are now responsible for five types of plan,
which deal respectively with long-term development,
any agglomerations within the region, its towns,
special areas such as those chosen for major
industrial or tourist growth, and each of the
communes. Commune, or local, plans are amongst the
most detailed of these five types, and must show the
proposed distribution of industry and warehouses,
housing of various densities, central and local
service centres, green areas, the basic
communication network, and areas in which land use
will not be allowed to change during the period of
the plan (Jachniak-Ganguly 1978, pp. 78-79, 97).

Decisions as to whether or not land-use change
shall take place are made at the different tiers in
the planning hierarchy, according to their
importance. For instance, the basic pattern of
economic and spatial development is determined by
the centre. All major industrial investments and
additions to infrastructure are agreed in the chief
planning body at the national level - the Planning
Commission of the Council of Ministers - and
decisions about them are sometimes the result of
bargaining at the very highest levels in the Polish
government (Nuti 1982, pp. 34-35). Put another
way, ninety percent of all investment is by

government, and approximately two-thirds of that in the socialised sector of the economy is controlled by the centre (Jachniak-Ganguly 1978, pp. 55, 63). Only the locations of developments of modest size and merely regional significance are determined by regional planning authorities, and even they must be fitted into the local-authority budgets, which are also controlled by central government. Nevertheless, regional authorities have been able to exercise their rights to refuse permission for nonconforming development by state enterprises after plans have been drawn up; and, for example, tar, benzole, coal-oxidation and coking plants, and large tanneries have all been refused permission for development in the city of Krakow for this reason (Jachniak-Ganguly 1978, pp. 68-73, 105-106).

Thus, land-use change appears to be very firmly in the control of central government in Poland, but it is not clear to what extent this control has been translated into effective land policies. At least until the late 1960s it appeared that land use was not given very great attention in comparison with the economic side of planning, and that location decisions were more concerned to maximise industrial output or to achieve a more even pattern of regional development than to meet any long-term aims for the best use of the country's land. Moreover, there is conflicting evidence as to whether or not the system of control is effective. A study of suburban towns to the north-east of Warsaw showed that over a ten-year period up to the mid-1970s less than five percent of the developments contravened the local plans (Chrobak 1976, pp. 12-14). But Michalak (1979, pp. 68-69) suggests that many of the second homes which have been built since 1970 in the mountains, and on the coast and the lakes, or close to the major cities, were not granted permission; and he has estimated that as many as two-thirds of those being built in the late 1970s, or about a thousand each year, were illegal.

Some idea of the type of land-use policies which exist at present in Poland may be obtained from the plans for one of the most important agricultural, industrial and tourist areas of the country, the south-east macro-region. The Spatial Development Plan for the whole country (Figure 7.3) provides the context for the plans which cover smaller areas, and it indicates amongst other things the areas in which mining, tourism and forestry are to be developed over a fifteen-year period. It also identifies the major growth centres in the country, and the areas which have good natural conditions for

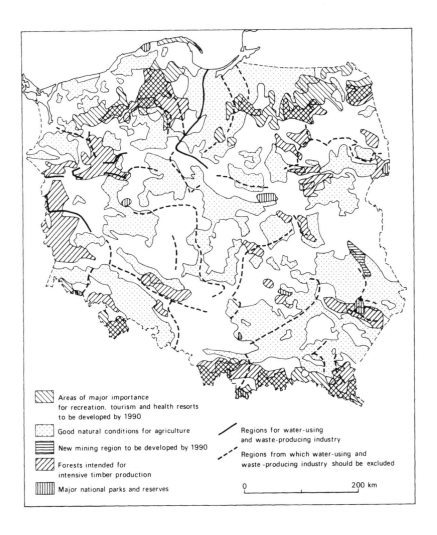

Areas of major importance
for recreation, tourism and health resorts
to be developed by 1990

Good natural conditions for agriculture

New mining region to be developed by 1990

Forests intended for
intensive timber production

Major national parks and reserves

Regions for water-using
and waste-producing industry

Regions from which water-using and
waste-producing industry should be excluded

0   200 km

Figure 7.3: Some Land-Use Elements of the Spatial
Development Plan for Poland

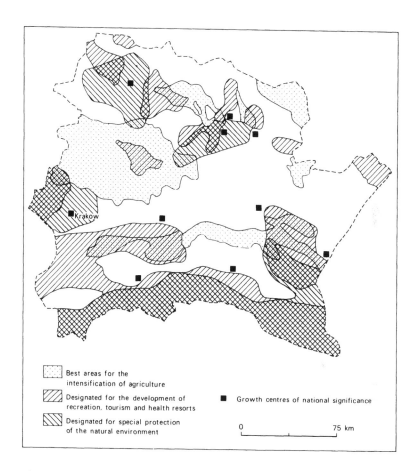

Best areas for the
intensification of agriculture

Designated for the development of
recreation, tourism and health resorts

Designated for special protection
of the natural environment

■ Growth centres of national significance

0   75 km

Figure 7.4: Some Land-Use Elements of the Development Plan
for the South-East Macro-Region of Poland

agriculture.   All   the   areas   shown   are   of   a
generalised   nature,   because   the   areal   scale   of   the
plan   is   small.   The   macro-region   plan   (Figure   7.4)
is   similar.   It   indicates   in   more   detail   the   best
farming   areas   and   those   designated   for   the
development   of   health   resorts,   tourism   and   the
protection   of   the   natural   environment,   but   there   is
considerable   overlap   between   the   zones,   and   their
boundaries   are   drawn   in   a   very   generalised,   even
diagrammatic,   manner,   and   are   not   related   in   detail
to topography or soil quality.

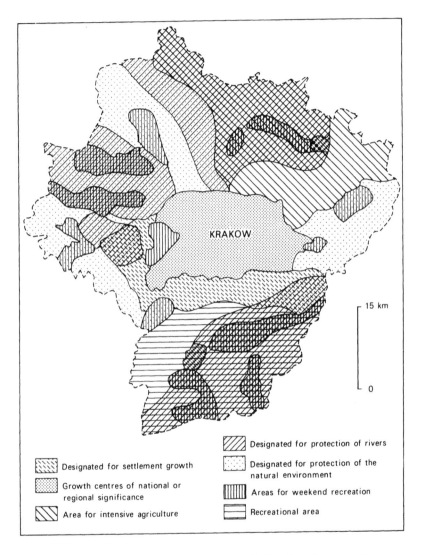

Figure 7.5: Some Land-Use Elements of the Development Plan
for Krakow Voivodship

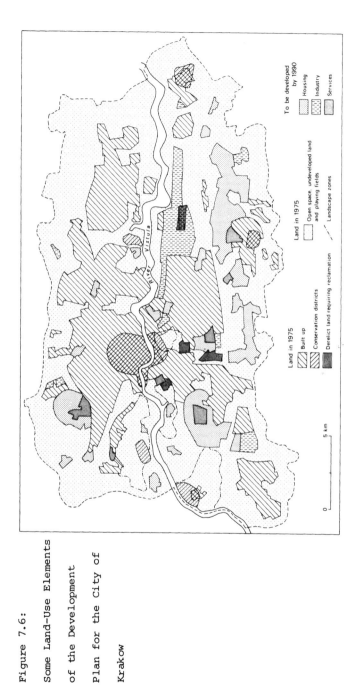

Figure 7.6:

Some Land-Use Elements

of the Development

Plan for the City of

Krakow

Table 7.2:   Some Details of the Perspective Plan for Krakow
             Voivodship and City to 1990

	Voivodship	
	1975	1990
Area (hectares)	325,400	325,400
Population	1,120,300	1,230-1,300,000
Percentage of Population in Towns	68.3	78
Land Use (hectares)		
Agriculture	228,400	209,200
Forest	56,000	59,700
Settlement and transport	29,400	47,300
Industry	3,000	4,100
Water	4,000	5,000

	City and Environs	
	1970	1990
Area (hectares)	47,000	47,000
Population	672,000	900-920,000
Built-Up Area (hectares)	9,980	28,169
of which		
Residential	3,814	7,560
Industry and warehousing	2,881	4,437
Transport and communications	1,633	6,065
Parks and sporting facilities	1,097	8,489
Services and administration	556	1,619

Sources: Aglomeracja Krakowska Plan Zagospodarowania
Przestrzennego Na Okres Perspektiwiczny, Krakow, p. 23;
Krakowski Zespol Miejski Plan Zagospodarowania
Przestrzennego Na Okres Perspektiwiczny, Krakow, p. 27.

Each of the regions within the macro-region is covered by its own spatial plan. That for the Krakow region envisages that the built-up area will increase very substantially between 1975 and 1990, almost all at the expense of farmland (Table 7.2) lying close to the city of Krakow (Figure 7.5). It also indicates that land has been set aside for a variety of conservation and recreational purposes, many of which overlap, but which nevertheless gives the local authorities power over the distribution of, say, private chalet and second-home development in scenic areas. Lastly, each commune is covered by a land-use plan, which shows the spatial arrangement of the developments which have been decided at national and regional levels, and the pattern of land-use zoning which any private developer will be obliged to follow. A comparison of the planned land-use totals for the city of Krakow and two neighbouring towns for 1990 with the actual figures for 1970 is given in Table 7.2. This shows that the proportion of the land area of the city which is to be developed is to rise three-fold over the period, in keeping with its status in the National Plan as a growth centre of national importance, with a sixty-three percent increase in the amount of residential accommodation per capita, a four-fold expansion of the area of transport facilities, and eight times as much land in parks and sporting facilities. Almost all of the land for these ambitious targets is to be found on the edges of the present built-up area (Figure 7.6), much of which is presently in agriculture, and is of good quality.

In addition to this general hierarchy of land-use policies there have been a number of more specific actions by the Polish government. One of these was the decision to rebuild the old city of Warsaw in its medieval form after its destruction during the war. Another was the Mining Act of 1953, which regulates the development of areas overlying mineral deposits. Thirdly, thirteen National Parks have been designated (Figure 7.3), largely under the Nature Protection Act of 1949, a further seven are planned, and several hundred smaller areas are protected from major land-use changes for cultural or scientific reasons. Much of this land is in public ownership, chiefly under the control of the Ministry of Forestry and Timber Industries, (Okolow 1980, pp. 4-6); other land within the Parks and reserves is being purchased by the state; and land around the Parks is included in their management plans where changes to it could directly affect the

nature of the designated parkland (Karpowicz 1982,
pp. 46-47). However, all this must be put in
context. Only about 0.6 percent of the country is
included in the Parks and reserves; not all of this
falls within their most strictly-protected cores,
from which all economic activity and development is
excluded (Glowny Urzad Statystyczny 1976, p. 168);
and much of it is managed with a view to timber
production rather than the protection of nature.
Indeed, it would appear that the Polish authorities
do not accord such 'protected' areas a very high
priority. In a discussion of the way in which the
Parks are dealt with in the regional plans Karpowicz
(1982, p. 47) says

"In the Ojcow National Park plan no account was
taken of the massive aerial pollution from the
Olkusz Region, whilst the Kampinos National
Park plan disregarded the pollution
possibilities of the Warsaw Steel Works and the
projected water regulation projects on the
nearby River Vistula. In the Pieninski
National Park plan the projected hydroelectric
dam and reservoir project on the River Dunajec
was opposed and alternative sites recommended.
However, due to the lack of specific and
detailed enough analysis and the lack of
sufficient funding from the national park
authorities, the protests only resulted in
minor alterations in the size, character and
form of the reservoir. No use was made of
regional and national plans in the search for a
better environmental solution. The result will
be increased investment in the immediate area
of the national park, greater tourist pressure
and uncontrolled landscape change."

However, by far the largest of the specific
actions of government to control land use, in terms
of the area affected, has been the legislation of
1971 concerning good quality agricultural land and
forests. This laid down that Classes I to IV
inclusive (out of a six-fold classification) of
agricultural land, and all forests, are to be
protected from development, and that industrial
users of land and the mining industry are obliged to
restore land which has been damaged. In doing this
the legislation applied to the whole country powers
which were already in existence in restricted
regions, and especially in Upper Silesia, where a
forest belt of 324 square kilometres had been
designated around the conurbation in 1968, and where

derelict land was already being restored (Brzezinski
1974, pp. 53-62). The 1971 legislation also
removed all agricultural land within city boundaries
which was earmarked for development under existing
plans from that category, and reduced the areas
which were designated for town extensions. In
addition, regional and local authorities were

Table 7.3:   Soil Class in Poland and Payments for the Transfer
of Land from Agriculture 1978

		Percent of all agricultural land	Payment (zlotys per hectare)
Class	I	0.4	500,000
Class	II	2.9	450,000
Class	III	21.6	350,000 - 400,000
Class	IV	39.6	150,000 - 300,000
Class	V	23.1	100,000
Class	VI	12.4	50,000
Lake Shores			300,000 - 600,000

Sources: Glowny Urzad Statystyczny, Rocznik Statystyczny
Wojewodztw 1979, Warszawa 1979, p. 210; Glowny
Urzad Statystyczny, Jakosc Uzytkow Rolnych,
Warszawa 1980a, p. 5; Hamilton, F.E.I., Spatial
Structure in East European Cities, in French, R.A.
and Hamilton, F.E.I. (eds.), The Socialist City,
Wiley, Chichester 1979, p. 209.

empowered to zone land of any quality around towns
with populations of fifty thousand or more for
agriculture or forest, and the zone over which such
control is permitted was extended for up to five
kilometres around any health or other resort. Thus
the large areas of rural Poland in which development
pressures from state-owned industry or private
housing might be expected to be particularly strong,
and two-thirds of all the arable land, have been
made subject to detailed land-use control, and all
proposals for the transfer of such land to other
uses must be approved by the local planning
authority. Moreover, where agricultural land is

released fees are payable by the developer which are
intended to cover the full cost of the transfer in
lost production for a period of twenty years (Table
7.3). However, national organisations, cultural,
health and social services, and state-owned housing
are not required to make these payments, leaving
only industry and mining which are usually called
upon to do so (Siuty 1978, pp. 253-256).

     Any assessment of the effect of this
legislation should acknowledge that, at least
initially, it appeared to fail in its aim of
protecting prime agricultural land. Between 1972
and 1975 sixteen percent of the Class I arable land
was transferred to other uses; and it was only after
the Ministry of Agriculture had issued an
Agricultural Code in 1974, which strengthened both
the controls over building in the countryside and
the policy, which had existed since 1961, of
concentrating rural house-building within villages,
that the rate of transfer fell (Czechowski 1977, p.
39; Glowny Urzad Statystyczny 1980a, p. 3). It may
be seen from Table 7.4, however, that, by the late
1970s, the proportion of the better agricultural
land which was transferred to other uses was tiny,
and that land of that quality was only being
transferred at about a fifth of the rate of the
poorest land, while forestry was experiencing a net
gain. On the other hand, only about half as much
land was being reclaimed, after being made derelict
by industry, mining or peat extraction, as was being
lost.

     However, it must be remembered that the
protection of rural land does not of itself
guarantee an adequate supply of agricultural or
forest products if other conditions surrounding
production are adverse. Thus, it is to other
aspects of the Polish economy than land that we must
look in order to explain the fact that, of all the
countries which have been examined in detail in this
study, it is the only one in which, in spite of the
measures taken to control the use of land, there was
not only a severe and overt food shortage in the
early 1980s, which led to the re-introduction of the
rationing of many items, but also a
continuingly-inadequate supply of land for housing,
as measured by the area of houseroom per capita.
For instance, in 1979, there was only 15.2 square
metres of houseroom per capita in the country
(Glowny Urzad Statystyczny 1980a, p. 272), and a
higher density of people per room than in all the
other countries of Eastern Europe, with the

Table 7.4:   Transfer of Land between Uses in Poland in 1978
             and 1979

	Total transferred (hectares)	Percent of total in country in 1978
Agricultural land		
Classes I to III	4574	0.1
Classes IV and V	15059	0.13
Class VI	11884	0.5
Forest Land	4088	0.05
Purpose of transfer		
Afforestation	7761	
Industry	6606	
Mining	3693	
Housing	7516	
Transport	2597	
Reservoirs and water supply	3012	
Others	4877	
Derelict land reclaimed	19646	19.1

Sources: Glowny Urzad Statystyczny, Rocznik Statystyczny
           Wojewodztw 1979, Warszawa 1979, pp. 210, 408-410;
           Glowny Urzad Statystyczny, Jakosc Uzytkow Rolnych,
           Warszawa 1980a, p. 3; Glowny Urzad Statystyczny,
           Rocznik Statystyczny Wojewodztw 1980, Warszawa
           1980b, pp. 444-446.

exception of Romania, or in Japan (Glowny Urzad
Statystyczny 1976, p. 353). Yet, as we have seen,
the quantity of land in Poland per capita is
considerably greater than in, say, Britain or Japan,
and is also more than that in Czechoslovakia, East
Germany and Hungary. Thus the problem is rather one
of the management of the land economy. Whereas in
Britain it is arguable that land-use intensity in
rural areas has been maintained at unjustifiably
high levels by the financial support which
governments have given to farming and forestry over
several decades, and that in Japan the government
has contrived a situation in which rural land is
underused while urban densities are too high, the
domestic deficits in farm and forest products in
those countries pose no problems of crisis
proportions, and are met by imports from other
countries with surplus production. However, the
problem in Poland seems to be altogether different.
Not only is the domestic production of food
inadequate, but also the level of imports, though
substantial and rising, does not appear to have been
sufficient to fill the gap; and, by the early 1980s,
that gap had grown so wide that it could not be
bridged, at least in the short run, by such
slowly-acting controls as those over land-use change
or atmospheric pollution.

The Polish government has sought other remedies
for this problem on several occasions over the
post-war period, and most recently in attempts to
raise the prices of food substantially in 1970, 1976
and 1980, each of which has led to serious unrest in
the country. These attempts - of which only the
last was successful - indicate one of the major
causes of the Polish problem, namely, the policy of
holding down food prices in the towns, and the
consequent slow rise in those paid to farmers, which
has severely limited the resources available for
investment on farms, and has encouraged younger
members of farming families to seek more highly-paid
employment in towns. This, in turn, has left an
ageing population, with a diminishing ability to
innovate, in charge of the farms. Since the early
1960s the state has offered to buy out such farmers
through the State Land Fund, but the majority of
those beyond retiring age have preferred to retain
control of their holdings, in view of the security
which it gives them. Furthermore, all these
developments which have had an adverse effect upon
farm output have been concentrated in the more
fertile south-east of the country, where most of the

land has remained in small, peasant holdings, where huge numbers of new industrial jobs have become available, and where farmers have been able to diversify into the provision of tourist or commuter accommodation, to the further neglect of their holdings (Dawson 1982, pp. 300-308).

CONCLUSION

We may conclude that Poland, like Japan, has only recently introduced an effective system of land-use control over most of the country. However, unlike all of the other countries in this study, it alone appears to face a severe land problem of a type which is more akin to those of developing countries, namely, a food shortage, which cannot apparently be met by imports, which are in turn paid for by exports of non-agricultural products. While it is likely that control over the transfer of good agricultural land to other uses may have some marginal effect upon the domestic output of farm goods in the long run, the contribution which land-use control can make to the solution of the country's immediate problem is slight, and the influence of market management and of the government's policy towards the private farmers is of much greater significance. A more balanced and freer market in farm products would probably lead to a substantial increase in their prices, and put an end to the huge disparity which grew up in the early 1980s between the official and black-market price levels; and this, in turn, would be likely to encourage higher levels of production. And it is especially necessary that the supply of, and demand for, farm products should be brought into balance in socialist economies, such as those in Eastern Europe, for artificially low or high levels of either cannot be offset by adjustments in the price of land in a situation in which land-use transfers in the socialised sector occur without payments for land in many cases, and where the market in private land is inhibited by the authorities and is subject to pre-emption by the state at fixed prices. It is true that, under the system of land-use and other controls which exists, Poland and the other economies of Eastern Europe do not suffer from some of the other serious land-use problems, such as inner-city decay, social segregation or suburban sprawl, which have appeared in many of the developed market economies, even where land-use controls have been in existence. But Poland does suffer from major land-related problems which have proved to be far more destabilising to the society as a whole than, for instance, even the inner-city troubles of Britain or the United States; and, to this extent, it would appear that its planners have failed to meet either of the two requirements made of them by Sampson in the quotation at the head of this chapter, namely, to identify the most appropriate criteria on which to base land-use decisions, or the best process by which those criteria may be brought to bear upon the land-use decision-making process.

Chapter 8

CONCLUSION

"In all countries there is a gap between the
accepted principles of land policy and the
means of implementation of these principles:
between the expressed principles of a programme
and the actual programme. What should be in
theory is amended or modified in practice to
appease interests." (N. Lichfield and H.
Darin-Drabkin, Land Policy in Planning, George
Allen and Unwin, London 1980, p. 194).

We began Chapter 1 by asking a series of
questions. We asked whether there is a land problem,
and if so, what form it takes. We enquired about
the adequacy or otherwise of the stock of land. We
raised the issue as to whether, if land is in
plentiful supply, problems may still arise because
of the way in which it is owned or used; and we
asked to what extent all these questions apply to
the developed economies. In Chapter 2 we examined
in more detail the nature of the land problem, first
as it has been seen by Malthus and his followers,
and secondly, as the result of the failure of the
free-market system; and we concluded that, while
developed economies are unlikely to face a
Malthusian threat, there are good reasons for
governments to intervene in the allocation of land
among alternative uses, and to control the spatial
distribution of those uses. However, it was noted
that intervention often allows a variety of actors
to participate in land-use decisions who would not
normally play a part if the matter were left to the
market, and who, in so far as they have contrasting
aims, resources and access to government, may exert
different degrees of influence upon such decisions.
The characteristics of each of a number of broad
types of such actors were described in Chapter 3.

Finally, the land problem was discussed in the cases
of four developed countries, each of which shares
some of the characteristics of the members of one of
the various sub-groups of developed economies which
were identified in the first chapter. This
discussion included a description of the supply of
land in each country, the recent pattern of demands
upon it, the attitude towards land ownership, and
the role of pressure groups in the system of
regulation, before giving an outline of the policy
and tools of land-use control which were in
operation in each during the early 1980s. The
discussion of each country was concluded with the
identification of what seemed to be some of the
chief problems arising out of that control. Thus,
much preparatory work has been done, but now we must
attempt to draw from it some general answers to the
questions which were posed at the outset, and to
assess the achievements of the countries which have
been studied, both against each other, and in
relation to the rest of the world, in tackling the
land problem.

In the first place it should now be quite clear
that there is a land problem in the developed
economies. Indeed, there seem to be many problems,
and that in spite of much regulation by the state.
Thus the Japanese have struggled with high land
prices, the British have worried about the
conservation of their countryside, Americans have
been involved in debates over both abandoned land in
the inner cities and the protection of rural,
wilderness areas, and some of the Eastern European
nations have faced serious problems of food supply,
to indicate but a few of the many aspects of the
land problem in these countries. All have debated
the question of the most appropriate pattern of land
ownership for the wellbeing of society as a whole.
But all this should come as no surprise, for the
stock of land in almost all developed economies, and
especially in settled areas, is almost everywhere
inadequate to meet the potential demand for it:
there are few places in which the price of land,
even excluding the value of improvements which have
been made to it, is zero. Gone are the days when it
was possible for a man to carve out as much of an
estate as he could master from the wilderness, after
the manner of the pioneer settlers in the American
West during the latter half of the nineteenth
century. What is more, even if there were
sufficient land for all who wanted it, we have seen
that the problem of externalities, and the more
general failure of the market to give adequate value

to the requirements of the community at large, as distinct from those of individual landowners, would very probably pose land problems, particularly in advanced economies. In short, very serious problems arising out of the overall and spatial allocation, and the intensity, of land use exist in some developed economies, which reflect the limited nature of the land area available to them.

But how serious are these problems in what are some of the richest countries in the world? In Chapter 1 we noted that the wealthy societies have the means whereby they can substitute other factors for land when it is in short supply, and can engage in trade to acquire the products of any economic activity for which they no longer have sufficient space. Nevertheless, land shortage does present a severe threat to some inhabitants of even the developed countries for, although such countries are not in general threatened by the possibility of absolute falls in their food supply per capita as population grows, as are many of the poorest nations, pollution of air, land and water adversely affects the health of many living in them. The careless dumping of wastes has reached levels in the period since 1945 at which it has caused deaths – through poisoned fish in Japan, respiratory diseases caused by smogs in Britain, and almost certainly by atmospheric pollution in southern Poland – and many less-serious, but unwanted, effects have been felt by large numbers of the populations of these countries, and also of the United States. Furthermore, high land prices have obliged the poor to live in cramped and congested environments even in some of the richest societies, have made them subject to unwanted displacement by redevelopment or gentrification, and have concentrated them in ghettoes of low-quality housing of mean environment, widely separated from the new locations of economic growth.

The immediate purposes for which governments in the four countries have taken action with respect to land appear to fall into three broad categories. Firstly, all have shown some wish to control the aggregate allocation of land among the alternative uses, perhaps with a view to securing a strategic reserve of, say, agricultural or forest land, or the protection of scenic or other valued environments. Secondly, all have wanted either to achieve particular levels of output, especially in farming, or to avoid congestion in urban areas, or both; and therefore have been concerned to influence the intensity with which land is used. Thirdly, they

all have perceived a need to control the
intermixture of uses in order to provide a safe and
pleasant environment, and thus to regulate their
areal pattern.
    The controls which have been adopted to achieve
these purposes may also be grouped into three:

    1. Indirect controls over specific sites or
    uses by fiscal measures
    2. Direct controls over specific sites or uses
    without the acquisition of land
    3. Direct control over specific sites or uses
    by means of land acquisition and its subsequent
    use by agencies of the state.

Those in the second group accord broadly, but not
entirely, with what is known as 'negative control',
while those in the third make up the 'positive'
planning of land use, and are the most powerful.
All of the countries in this study have made use of
all three types of control in the recent past, but
there have been wide variations amongst them in the
emphasis which has been placed on one as against
another. Thus the Eastern European countries have
made most use of land acquisition, either directly
or through the collectivisation of agriculture, and
in Britain the authorities have also purchased much
land for a variety of purposes, whereas, as least in
this century, land acquisition by the American and
Japanese governments has been small. Similarly,
while some level of government in Britain, Japan and
Poland has detailed plans of the pattern of uses
which it would like to see, and while such direct
controls over particular sites have been the
mainstay of British land-use regulation, much of the
United States remains unzoned. In contrast, there
has been much use of indirect control in all the
countries, either in the form of subsidies for
agriculture in the market economies, or its taxation
in Poland; and charges for the use of land by some
organisations in Eastern Europe, and variations in
the level of the property tax in the United States
in return for the preservation of particular uses on
specified sites, also show that similar methods of
control have been adopted in countries which
otherwise have very different ways of running their
economies.
    Comparison of the land-use regulations in the
four countries suggests that, in addition to any
indirect financial influences which governments can
bring to bear upon land-use decisions, Japan and
Poland now possess more wide-ranging controls,

affecting a larger proportion of the land in those
countries, than do the United States or, perhaps,
Britain. On the other hand, it should be remembered
that in Japan, as in other market economies, most
land is in private ownership, and that therefore
many controls can only be used in a negative manner,
whereas in most of the centrally-planned economies
most land is communally owned. Moreover, as we
noted in Chapter 5, it does not appear as though
Japan is making full use of the controls which have
been created. Paradoxically, the United States'
federal government, through its residual ownership
of land, is in a position to control directly almost
as high a proportion of that country as the Polish
government can of its.

But do these controls amount to a land policy,
and have they been drawn up in conjunction with an
explicit and coherent land budget, in which targets
have been set for individual land uses in the
future? To this question the answer must be 'no'.
None of the four countries have a comprehensive land
policy as such, though all have policies concerning
separate parts or uses of their land. Even Japan's
National Plan, which includes some land-use
forecasts, has not been backed by sufficiently-firm
direction from central government for the changes in
use to be kept in line with the Plan, while Britain,
for all its recent emphasis on structure planning
and control at local level, has no targets for the
country as a whole on which local-authority plans
could be based. Instead, planning bodies in Britain
are offered general criteria which they are expected
to bear in mind when coming to decisions, while
central government retains extensive powers to alter
those decisions when it does not approve of them,
with the result that government intervention is
characterised by great uncertainty and frequent
controversy. Similarly, in the United States, where
there is also a lack of detailed criteria from the
federal government to guide decision-making
authorities, a wide variety of both type and degree
of controls exists between one part of the country
and another, and the extensive use of litigation
introduces further uncertainties into the system of
land-use regulation. However, there may be little
purpose in setting targets for the use of land when
they will inevitably require modification as
economic circumstances change. For example, both
Japan and Poland have found since the Second World
War that economic plans - on which land-use targets
might be based - are rarely achieved exactly in the
form in which they are originally drawn up; and

Britain's only attempt at an economic plan soon
collapsed in the mid-1960s. Furthermore, advances
in technology redefine the resource base of any
country, and lead to the re-evaluation of its land,
both by landowners and by society at large; and it
is in the developed economies that those advances
are most likely to occur.

The success or otherwise of the policies which
do affect land use may be assessed by relating them
to the basic forms of the land problem, and in
particular to the extent to which they have

1. obliged land users to meet the full costs
of their activities;
2. led to the full reward of owners for the
use which is made of their land by other
people;
3. been able to recoup for society as a whole
the gains in the value of land which result not
from the owner's efforts, but from social
action, while still allowing land to be put to
its most profitable use; and
4. secured the needs of society as a whole in
matters of land use, where these are
incompatible with the working of the private
market in land.

When assessed against these criteria land-use
regulation in the four countries does not appear to
have been very successful. For example, none of the
four has succeeded in obliging land users to meet
the full costs of their activities, and this is
particularly true in Poland, where industrial
pollution continues to be a serious problem, and in
market economies, where the costs of clearing away
abandoned housing, mills and industrial dereliction
still falls in part upon the state. Moreover, while
the usual method which has been employed to deal
with nonconforming uses - zoning - may have been
appropriate to the solution of very localised
spillover problems, such as noise or road safety, it
has proved to be quite incapable of solving the
problems of long-distance atmospheric pollution from
tall chimneys, as have formal prohibitions against
the emission of pollutants. Furthermore, zoning may
have prevented land users from bidding for the sites
which would be most profitable for them, even after
they have met the full costs associated with their
activities, and thus may have reduced the total of
rents produced from all the sites in a country.
Nevertheless, the need to oblige land users to meet
the full costs of their activties has been accepted

in some, if not in all, cases in most of the
developed countries.

Similarly, there is a general acceptance in the
market economies that owners who are restricted in
the way in which they may use their land, in order
to ensure that it is of continuing attraction or
interest to others, should be compensated. This
applies in particular to the United States, but it
is also the case in Britain, where restrictions on
the improvement of land by farmers may still entitle
those farmers to payments from public funds.
However, there has been a trend for the communal
rights in land to be extended, especially during the
1960s and early 1970s, with the effect that many
aspects of land which were previously considered to
be a part of the property of the owner have now, in
effect, become public property; and in Poland this
has been carried to a point at which almost all
recognised scenic land is in public ownership. Thus
the problem of compensation is diminished and, in
some instances, avoided altogether.

Conversely, the third aim of government
intervention in land-use matters is by no means
universally accepted among the developed economies,
even where planning occurs. Neither in the United
States, nor in Japan, is betterment recouped for the
community; in Britain it is only recovered in part;
and in Poland the variable, if low, levels of
compensation which are paid to landowners, depending
on their location, for the acquisition of their
property, suggests that this aim may not be fully
implemented even in socialist economies. One
practical reason for the failure to recover such
gains to landowners is the marked tendency, in the
absence of a system of land prices which reflect the
potential value of sites to a range of alternative
users, for the hoarding of land, and its inefficient
use. On the other hand, the failure to recoup that
part of the land value which has been created by
society at large in those economies in which
land-use regulation occurs merely encourages
speculative purchases and increases the windfall
gains to those owners who are lucky enough to be
able to get permission to develop their land or to
change its use to something more profitable. It
also encourages owners to attempt to influence the
decisions of the regulatory authorities.

Thus there is a wide variety of performance
among the four countries in relation to the first
three aims of land-use regulation. However, it is
in the nature of modern government, especially at
national level, to espouse the fourth, and to secure

at least some of the needs of society with regard to
land for which the market, on its own, might not
have provided. In particular, land for defense,
recreation and transport, the protection of adequate
areas for agricultural, forest and mineral
production, and the provision of sufficient space to
enable housing to be built at acceptable densities
for the poor have all been accepted as the
responsibility of central government in each of the
four countries which have been covered in this
study. This is not to say that all governments have
accepted these responsibilities to the same extent,
nor that all have adopted the same means by which to
exercise them, and least of all to suggest that all
have been equally successful in carrying out
policies designed to combat the effects of market
failure. For example, the limited and indirect
attempts by the federal authorities in the United
States to encourage the provision of low and
moderate-income housing stands in sharp contrast to
the directly-funded and centrally-planned
public-housing programme in Poland. Nevertheless,
the aim of government - to provide housing of at
least some minimum standard and maximum density, and
thus to influence the allocation of land among
competing uses - is similar.

     Thus we may conclude that the problems which
have arisen from the shortage of land in the
developed economies have been tackled only in part,
and that therefore the landowner or user - be he a
person, firm or an agent of government - continues
to enjoy significant advantages over other members
of the community, and that this conclusion reflects
the scope which landowners and users enjoy to
influence planners and governments even in
centrally-planned, 'socialist' economies.

     But what does all this amount to in the end of
the day? When all the problems have been identified,
and the efforts of government to tackle them have
been described, what has been the outcome of the
struggles with the land problem? One method of
comparing the effect of regulation might be to
examine the quality of life which each of the
various countries has come to enjoy in the period
since controls were adopted. However, such a basis
of comparison should be treated with care. Not only
do patterns of personal preference and demand vary
between societies with similar levels of wealth, but
also, although all the countries which have been
studied are classified as developed economies, there
are great differences in the degree of development
between, say, the United States and Poland.

Conversely,   it should be remembered that Japan and
Poland had similar economic structures, with large
proportions   of   their   populations   in   peasant
agriculture, at the end of the Second World War, and
gross national products per capita which were very
much   less   than   those   in   Britain   or   the United
States. Furthermore, sufficient time has elapsed to

Table 8.1:   Selected Indices of the Quality of Life

	Japan	Poland	United Kingdom	United States
Food supply				
Calories per capita per diem in 1978-80	2916	3520	3316	3652
Protein in grams per capita per diem in 1978-80	93.4	105.4	91.4	106.7
Meat and fish (kilograms per capita per annum in 1978)	N.A.	78	78[1]	117[2]
Cereals and potatoes (kilograms per capita per annum in 1978)	N.A.	288	159[1]	102[2]
Housing (people per room in 1978)	1.1	1.2	0.6[3]	0.6[4]

1.   in 1976/77
2.   in 1975
3.   in 1971
4.   in 1970

Sources: FAO Production Yearbook, Rome 1982, Tables 97-99;
           Glowny Urzad Statystyczny, Rocznik Statystyczny
           1980, Warszawa 1980, p. 571.

allow   different rates of economic growth to close
the gap which existed between Japan and Britain in
many of the characteristics of development which

were used in Table 1.1. Therefore, it may not be
entirely misleading to judge the success or
otherwise of land-related policies by making
comparisons of the supply of some of the basic
necessities which are closely connected with land in
the four countries; and these are set out in Table
8.1. While these figures do not tell us very much
directly about that success, they do indicate, for
example, that levels of nutrition are more than
adequate in all the countries, although they were
falling in the late 1970s and early 1980s in Poland;
that, notwithstanding its apparent shortage of land,
Japan has managed to provide similar quantities of
both houseroom and food (from home production and
imports) for its people as has Poland; and that
Britain, in spite of a much higher density of
population, may have been almost as successful in
the provision of housing as the United States.

Lastly, some attempt should be made to relate
the experiences of the four countries to the rest of
the world. In the first place, it is hoped that
each of them represents to some extent one or other
of the sub-groups of developed economies which were
identified in Chapter 1 (Table 1.2), and thus may
stand not only for itself but also for a number of
countries with similar land resources, attitudes
towards land, and histories of tackling the land
problem. For example, Britain belongs to the group
of densely-populated and highly-developed countries,
which includes the Netherlands and West Germany
amongst others, both of which also have a
long-established tradition of land-use control
(McKay 1982; Sutcliffe 1981, pp. 9-46; United
Nations 1973, pp. 149-151); while Japan may be
illustrative of the way in which market economies
which have enjoyed rapid growth in the recent past,
such as Italy and Taiwan, have begun to grapple with
the land problem (Fuchs and Street 1980; Merlo
1979). Similarly, the pattern of rather weak and
patchy land-use control in the United States,
stemming from its history of recent settlement and
huge land resources, is also to be found in
Australia, the developed countries of South America,
and, to a lesser extent, in Canada (United Nations
1975, pp. 184-187); while Poland's problems are
typical of those of much of Eastern Europe.

However, it must also be acknowledged that the
effects of land-use regulation, or the lack of it,
are not confined within frontiers; and attention has
already been drawn in earlier chapters to two major
problems of this type which remain to be solved.
One of these arises out of the long-distance

atmospheric pollution which crosses national
boundaries, and to which Poland and Britain have
both been contributing in large measure; but which
also manifests itself in the way in which some
countries, and especially Poland, may not have been
as diligent in protecting those of their monuments
which have been recognised as being of global
significance as they might from the damage caused by
such pollution. The second is that to which Isenman
and Singer (1977, pp. 205-237), McGovern (1979, p.
8) and many others have drawn attention, namely, the
way in which the policy of protecting agriculture,
which is practised in all three of the market
economies, and especially in Japan, may be reducing
the markets which would otherwise be available to
producers in some of the poorer countries, and also
giving rise to a need to export the subsidised,
surplus farm goods at low prices to the Third World,
with the effect of further depressing the levels of
agricultural output there. More effective control
of these practices is required, perhaps through the
United Nations' Food and Agricultural Organisation
(FAO) or GATT, but success in this will depend to a
large extent upon the ability of governments to
resist the demands of their farm lobbies. At the
same time, the heavy dependence of some Third World
countries upon food imports from the United States
and other developed countries, which is expected to
continue well into next century, suggests that there
is a need to protect the potential for production,
in the form of land, in those economies, if not the
farmers who are currently producing. If this does
not occur, and if the supply of food to the
developing countries should become even more
inadequate in the near future, some at least of the
blame will fall upon the developed world. In short,
at the global scale, just as at those of the nation
and the local community, the land problem in the
developed economy has yet to be solved.

BIBLIOGRAPHY

Abercrombie, P. (1945) Greater London Plan, HMSO, London

Aglomeracja Krakowska Plan Zagospodarowania Przestrzennego Na Okres Perspektiwiczny (Undated), Krakow

Agricultural Economic Development Council (1977) Agriculture into the 1980s: Land Use, National Economic Development Office, London

Allen, G.C. (1974) 'The Causes of Japan's Economic Progress' in C. Nishiyama and G.C. Allen (eds.), The Price of Prosperity, Hobart Paper 58, Institute of Economic Affairs, London, pp. 31-62

Ambrose, P. and Colenutt, B. (1975) The Property Machine, Penguin, Harmondsworth

Ames, E.A. (1981) 'Philanthropy and the environmental movement in the United States', Environmentalist, 1, 9-14

Anderson, M.M. and Harper, J.P. (1978) Land Use and National Coal Surface Mining Policies, Argonne National Laboratory, Illinois

Andrews, R.N.L. (ed.) (1979) Land in America, Lexington Books, Lexington

Baerwald, T.J. (1981) 'The site selection process of suburban residential builders', Urban Geography, 2, 339-357

Bale, M.D. and Lutz, E. (1981) 'Price distortions in agriculture and their effects: an international comparison, American Journal of Agricultural Economics, 63, 8-22

Barna, T. (1979) Agriculture towards the year 2000: Population and trade in high income countries, Sussex European Research Centre, University of Sussex

Barlow, (1940) Report of the Royal Commission on the Distribution of Industrial Population, Cmd 6153, HMSO, London

Bater, J.H. (1977) 'Soviet town planning: theory and practice in the 1970s', Progress in Human Geography,

1, 177-207

Batie, S.G. and Healy, R.G. (1980) *The Future of American Agriculture as a Strategic Resource*, The Conservation Foundation, Washington

Beckerman, W. (1974) *In Defence of Economic Growth*, Jonathan Cape, London

Berque, A. (1980) 'La Montagne et L'Oecoumène au Japon', *L'Espace Geographique*, 9, 151-162

Berry, D. (1978) 'Effects of Urbanisation on Agricultural Activities', *Growth and Change*, 9, 2-8

Berry, J.K., Parker, J.K. and Burch, W. (1982) 'Rural lands in the USA' *Countryside Planning Yearbook*, 206-217

Best, R.H. (1981) *Land Use and Living Space*, Methuen, London

Bjork, B.C. (1980) *Life, Liberty and Property*, Lexington Books, Lexington

Blacksell, M. and Gilg, A. (1981) *The Countryside: Planning and Change*, George Allen and Unwin, London

Boddington, M. (1973) 'A food factory', *Built Environment*, 2, 443-445

Bosselman, F.P. and Callies, D. (1971) *The Quiet Revolution in Land Use Controls*, United States Council of Environmental Quality, Washington D.C.

Boyer, M.C. (1981) 'National Land Use Policy: Instrument and Product of the Economic Cycle' in J.I. de Neufville (ed.), *The Land Use Policy Debate in the United States*, Plenum Press, New York, pp. 109-125

Brotherton, D.I. (1983) *Ministerial Appointments to National Parks*, Council for National Parks, London

Brown, L.R. (1978) *The Worldwide Loss of Cropland*, Worldwatch Institute, Washington D.C.

Brown, L.R. (1980) *Food or Fuel: New Competition for the World's Cropland*, Worldwatch Institute, Washington D.C.

Brown, T.L., Miller, D.J. and Wilkins, B.T. (1977)

'Rural Nonfarmed Lands and their Owners in Five Central New York Counties', Search, 7, 1-23

Brzezinski, W. (1974) Legal Protection of Natural Environment in Poland, Panstwowe Wydawnictwo Naukowe, Wroclaw

Buringh, P., Van Heemst, H.D.J. and Staring, G.J. (1975) Computation of the absolute maximum food production of the world, Agricultural University, Wageningen

Callies, D. (1981) 'Public Participation in the United States, Town Planning Review, 52, 286-296

Centre for Agricultural Strategy (1980) Strategy for the UK Forest Industry, Report No. 6, University of Reading

Chadwick, E. (1888) The Malthusian Theory, East Sheen

Cherry, G. (1982) The Politics of Town Planning, Longman, London

Chou, M., Harmon, D.P., Kahn, H. and Wittmer, S.H. (1977) World Food Prospects and Agricultural Potential, Praeger, New York

Chow, W.T., 'Planning for micropolitan growth in California, Town Planning Review, 52, 184-204

Chrobak, A. (1976) 'Anglo-Polish Land Use Project - The Warsaw Survey Area', typescript, Warsaw

Clark, G. (1982) 'Beauty areas may not be given protected status', The Times, London, 3 July, 2

Clayton, H. (1982) 'Farm aid "at expense of wildlife"', The Times, London, 21 October, 4.

Coleman, A. (1976) 'Is planning really necessary?', Geographical Journal, 142, 411-437

Coleman, A. (1978a) 'Agricultural land losses: the evidence from maps' in A.W. Rogers (ed.), Urban Growth, Farmland Losses and Planning, Wye College, London, pp. 16-36

Coleman, A. (1978b) 'Last bid for land-use sanity', Geographical Magazine, 50, 820-824

Coughlin, R.F. (1979) <u>Agricultural Land Conversion in the Urban-Rural Fringe</u>, Regional Science Institute, Philadelphia

Coughlin, R.F. and Berry, D. (1977) <u>Saving the Garden: The Preservation of Farmland and other Environmentally Valuable Land</u>, Regional Science Research Institute, Philadelphia

Council of State Governments (1982) <u>Forest Resource Management: Meeting the Challenge in the States</u>, Lexington

Coyle, W.T. (1981) <u>Japan's Rice Policy</u>, USDA Economics and Statistics Service, Foreign Agricultural Economic Report No. 164, Washington D.C.

Culhane, P.J. (1981) <u>Public lands politics: interest groups' influence on the Forest Service and the Bureau of Land Management</u>, Resources for the Future Inc., Baltimore

Cullingworth, J.B. (1979) <u>Town and Country Planning in Britain</u>, 7th ed, George Allen and Unwin, London

Czechowski, P. (1977) 'Wplyw Rozwiazan Prawnych na Przestrzenne Rozmieszczenie Terenow Budowlanych na Obszarach Wsi' in <u>Wplyw Instrumentow Prawnych na Przestrzenna Strukture Rolnictwa w Polsce</u>, Polska Akademia Nauk, Komitet Przestrzennego Zagospodarowania Kraju, Warszawa

Darke, R.A. (1979) <u>Monitoring the Structure Planning Process within a new Metropolitan County Council</u>, Social Science Research Council, Report No. HR 3363, London

Dawson, A.H. (1971) 'Urban Structure in Free-Enterprise and Planned Socialist Environments', <u>Tijdschrift voor Economische en Sociale Geographie</u>, <u>52</u>, 104-113

Dawson, A.H. (1982) 'An Assessment of Poland's Agricultural Resources', <u>Geography</u>, <u>67</u>, 297-309

de Neufville, J.I., ed. (1981) <u>The Land Use Policy Debate in the United States</u>, Plenum Press, New York

Dempster, P. (1969) <u>Japan Advances</u>, 2nd edn., Methuen, London

Department of the Environment (1974) <u>Circular 98/74</u>, London

Department of the Environment (1978) <u>Land Availability: A Study of Land with Residential Planning Permission</u>, Economist Intelligence Unit, London

Dobby, A. (1978) <u>Conservation and Planning</u>, Hutchinson, London

Donnelly, M.W. (1977) 'Setting the Price of Rice: A Study in Political Decision making' in T.J. Pempel (ed.), <u>Policymaking in Contemporary Japan</u>, Cornell University Press, Ithaca, pp. 143-200

Eberhardt, P. (1980) 'Zarys wybranych metod z ekonomiki planowania ukladow osadniczych', <u>Przeglad Geograficzny</u>, <u>52</u>, 519-541

Eckholm, E.P. (1976) <u>Losing Ground: Environmental stress and world food prospects</u>, Norton, New York

Edel, M. (1981) 'Land Policy, Economic Cycles, and Social Conflict' in J.I. de Neufville (ed.), <u>The Land Use Policy Debate in the United States</u>, Plenum Press, New York, pp. 127-139

Edwards, A.M. and Wibberley, G.P. (1971) <u>An Agricultural Land Budget for Britain 1965-2000</u>, Wye College, London

Ehrlich, P.R. and Ehrlich, A.H. (1970) <u>Population Resources Environment. Issues in Human Ecology</u>, W.H. Freeman & Co., San Francisco

Environmental Agency (1980) <u>The Quality of the Environment in Japan</u>, Tokyo

Fabry, V. (1979) 'Protection of Agricultural Land', <u>Bulletin of Czechoslovak Law</u>, <u>18</u>, 82-91

FAO (1972) <u>Production Yearbook 1971</u>, Rome

Ford, L.R. (1979) 'Urban preservation and the geography of the city in the USA', <u>Progress in Human Geography</u>, <u>3</u>, 211-238

French, R.A. and Hamilton, F.E.I., eds. (1979) <u>The Socialist City</u>, Wiley, Chichester

Fuchs, R.J. and Street, J.M. (1980) 'Land

Constraints and Development Planning in Taiwan', _The Journal of Developing Areas_, 14, 313-326

Fuguitt, G.U. and Voss, P.R. (1979) _Growth and Change in Rural America_, Urban Land Institute, Washington D.C.

Furuseth, O.J. (1979) 'The Structure of Agricultural Land Conversion in Washington County, Oregon, _Journal of Soil and Water Conservation_, 34, 138-141

Furuseth, O.J. (1981) 'Update on Oregon's Agricultural Protection Program: A Land Use Perspective', _Natural Resources Journal_, 21, 57-70.

Gaidamara, E. (1981) 'Possibilities for the better use of land', _Planovoe Khozyaistwo_, 5, 519-541 (in Russian)

Ganguly, D. (1976) 'General Review of Planning Administration and Land Management in Poland and Britain', typescript, East Anglia University

George, H. (1979) _Progress and Poverty_, Robert Schalkenbach Foundation, centenary edn., New York

Gilbert, W. and Gregor, D. (1973) _Minnesota Land Use Laws_, University of Minnesota, Minneapolis

Ginsberg-Gebert, A. (1981) 'O samorzadnosc gospodarki miejskiej', _Miasto_, 31/4, 5-11

Glasgow District Council (1978) _Dalmarnock Local Plan, Written Statement_, Glasgow

Glickman, N.J. (1979) _The Growth and Management of the Japanese Urban System_, Academic Press, New York

Glowny Urzad Statystyczny (1976) _Maly Rocznik Statystczny 1976_, Warszawa

Glowny Urzad Statystyczny (1979) _Rocznik Statystyczny 1979_, Warszawa

Glowny Urzad Statystyczny (1980a) _Gospodarka Zywnosciowa 1980_, Warszawa

Glowny Urzad Statystyczny (1980b) _Jakosc Uzytkow Rolnych_, Warszawa

Glowny Urzad Statystyczny (1980c) _Rocznik Statystyczny Wojewodztw 1980_, Warszawa

Goldsmith, M. (1983) Politics, Planning and the City, Hutchinson, London

Grigg, D. (1981) 'The historiography of hunger: changing views on the world food problem 1945-1980', Transactions of the Institute of British Geographers, 6, 279-292

Grocholska, J., ed. (1972) Bilans Uzytkowania Ziemi w Polsce, Polska Akademia Nauk, Instytut Geografii, Warszawa

Guither, H.D. (1980) The Food Lobbyists: Behind the Scenes of Food and Agri-Politics, Lexington Books, Lexington

Hall, P. (1977) The World Cities, 2nd edn., Wiedenfeld and Nicolson, London

Hamilton, F.E.I. (1979) 'Urbanisation in Socialist Eastern Europe: the Macro-Environment of the Internal City Structure' in R.A. French and F.E.I. Hamilton (eds.), The Socialist City Wiley, Chichester, pp. 167-194

Hamilton, F.E.I. and Burnett, A.D. (1979) 'Social Processes and Residential Structure' in R.A. French and F.E.I. Hamilton (eds.), The Socialist City, Wiley, Chichester, pp. 263-304

Harrison, F. (1983) The Power in the Land, Shepheard-Walwyn, London

Hart, J.F. (1976) 'Urban Encroachment on Rural Areas', Geographical Review, 66, 1-17

Hase, T. (1981) 'Japan's Growing Environmental Movement', Environment, 23, 14-36

Hathaway, D.E. (1981) 'Government and agriculture revisited: a review of two decades of change', American Journal of Agricultural Economics, 63, 779-787

Hayakawa, K. (1978) 'Housing and the quality of life', Built Environment Quarterly, 4, 23-26

Healy, R.G. (1979) 'Land Use and the States: A Variety of Discontents' in R.N.L. Andrews (ed.), Land in America, Lexington Books, Lexington, pp. 7-21

Healy, R.G. and Rosenberg, J.S. (1979) Land Use and the States, 2nd edn., Resources for the Future Inc., Baltimore

Herington, J. (1982) 'Circular 22/80 - the demise of settlement planning', Area, 14, 157-166

HMSO (1944) The Control of Land Use, London

HMSO (1975) Food from our own Resources, Cmnd 6020, London

HMSO (1979) Farming and the Nation, Cmnd 7458, London

Honey, R. and Eriksen, R.H. (1979) Locational Equity, Land Use, and Minnesota's Fiscal Disparity Act, University of Iowa, Iowa City

House of Commons (1972) Report of Select Committee on Scottish Affairs, Paper 511, HMSO, London

Hoyle, F. (1963) A Contradiction in the Argument of Malthus, University of Hull

Irland, L.C. (1979) Wilderness Economics and Policy, Lexington Books, Lexington

Isenman, P.J. and Singer, H.W. (1977) 'Food Aid: Disincentive Effects and their Policy Implications', Economic Development and Cultural Change, 25, 205-237

Itakura, K. (1980) 'Cycles of Industrial Employment Agglomeration in Japan', Science Reports, Tohoku University, 7th series, Geography, 30, 83-95

Jachniak-Ganguly, D. (1978) Administration and Spatial Planning as Tools of Land Management in Poland, Centre for Environmental Studies, Occasional Paper No. 4, London

Jackson, R.H. (1981) Land Use in America, V.H. Winston and Sons, London

Jastrzebski, S. (1976) Kierunki ochronny srodowiska przyrodniczego w Polsce, Polska Akademia Nauk, Wydzial Nauk Rolniczych i Lesnych, Warszawa

Kahn, H. (1976) The Next 200 Years: A Scenario for America and the World, Morrow, New York

Kakiuchi, G.H. and Hasegawa, M. (1979) 'Recent
Trends in Rural to Urban Migration in Japan: The
Problem of Depopulation', Science Reports, Tohoku
University, 7th series, Geography, 29, 47-61

Kaneyasu, I. (1981) 'Introduction to Landscape
Assessment (1)', Science Reports, Tohoku University,
7th series, Geography, 31, 40-48

Kansky, K.J. (1976) Urbanization under Socialism,
the Case of Czechoslovakia, Praeger, New York

Karpowicz, Z. (1982) 'Conflict Solving in National
Park Management Plans: Examples from Poland' in
Essays in National Resource Planning, Centre for
Urban and Regional Studies Occasional Paper, No. 5,
University of Birmingham, pp. 37-57

Kirk, G. (1980) Urban Planning in a Capitalist
Society, Croom Helm, London

Knox, P. and Cullen, J. (1981) 'Planners as urban
managers: an exploration of the attitudes and
self-image of senior British planners', Environment
and Planning A, 13, 885-898

Kondratieff, N.D. (1935) 'The long waves in economic
life', Review of Economics and Statistics, 17,
105-115

Krakowski Zespol Miejski Plan Zagospodarowania
Przestrzennego Na Okres Perspecktiwiczny (undated),
Krakow

Leszczycki, S. (1980) 'Granice wzrostu ludnosci
swiata', Acta Universitatis Carolinae, Geographia,
15, 95-99

Lenin, V.I. (1961) Polnoye Sobraniye Sochineniy,
vol. 26, Gospolitizdat Press, Moscow

Ley, D. and Mercer, J. (1980) 'Locational conflict
and the politics of consumption', Economic
Geography, 56, 89-109

Lichfield, N. and Darin-Drabkin, H. (1980) Land
Policy in Planning, George Allen and Unwin, London

Little Rock (1980) Suburban Development Plan, Little
Rock

Lowe, P.D. and Wibberley, G.P. (1981) A Political Analysis of British Rural Conservation Issues and Policies, Social Science Research Council Report HA 5010, London

Lyons, M.S. and Taylor, M.W. (1981) 'Farm Politics in Transition: the House Agriculture Committee', Agricultural History, 55, 128-146

McAuslan, P. 'Local government and resource allocation in England: changing ideology, unchanging law', Urban Law and Policy, 4, 215-268

McConnell, W.P. (1975) Remote Sensing 20 Years of Change in Massachusetts 1951/2 - 1971/2, Massachusetts Agricultural Experiment Station, Research Bulletin No. 630, Amherst

McGlen, N.E., Milbrath, L.W. and Yoshii, H. (1979) 'Cultural Differences in Perceptions of Environmental Problems', Technological Forecasting and Social Change, 14, 97-114

McGovern, G. (1979) 'Human Rights and World Hunger', National Forum, 69, 8

McKay, D.H., ed. (1982) Planning and Politics in Western Europe, Macmillan, London

McKean, M.A. (1977) 'Pollution and Policymaking' in T.J. Pempel (ed.), Policymaking in Contemporary Japan, Cornell University, Ithaca, pp. 201-238

McKeown, T. (1976) The Modern Rise of Population, Edward Arnold, London

McNelly, T. (1972) Politics and Government in Japan, 2nd edn., Houghton Mifflin Co., Boston

Malin, K. (1976) How Many will the Earth Feed?, Progress Publishers, Moscow

Malthus, T.R. (1798) Essay on the Principle of Population as it affects the future Improvement of Society, J. Johnson, London

Marx, K. and Engels, F. (1959) 'The Manifesto of the Communist Party' in L. Feuer (ed.), Basic Writings on Politics and Philosophy: Karl Marx and Friedrich Engels, Doubleday, Garden City

Massachusetts (1982) The Zoning Act, Massachusetts

General Laws, Chapter 40A

Massey, D. and Catalano, A. (1978) _Capital and Land_, Edward Arnold, London

Matsui, T., ed. (1976) _IFHP Bulletin 1976, Japan - special number_

Meadows, D.H. and Meadows, D.I. (1972) _The Limits to Growth: a report for the Club of Rome's project on the predicament of mankind_, Universe Books, New York

Meek, R.L. (1953) _Marx and Engels on Malthus_, Lawrence and Wishart, London

Mera, K. (1977) 'The Changing Pattern of Population Distribution in Japan and its Implications for Developing Countries', _Habitat International_, 2, 455-479

Merlo, M. (1979) 'Anglo-Italian Land Use Policies', unpublished Ph.D. thesis, Wye College, London

Mesarovic, M.D. and Pestel, E. (1975) _Mankind at the Turning Point, the second report to the Club of Rome_, Hutchinson, London

Michalak, J. (1979) 'Indywidualne domki rekreacyjne na wsi', _Wies Wspolczesna_, 23, 67-74

Mills, E.S. and Song, B. (1977) _Korea's Urbanisation and Urban Problems_, Harvard University Press, Cambridge

Ministry of Agriculture, Fisheries and Food (1968) _A Century of Agricultural Statistics_, HMSO, London

Ministry of Agriculture, Forestry and Fisheries (1980) _Long-Term Prospects for the Demand and Production of Agricultural Products_, Tokyo

Minnesota (1978) _Notebook on Land Use Projections_, State Planning Agency, Environmental Planning Division, St. Paul

Minnesota (1981) _Growth Management Study_, State Planning Agency, Physical Planning Division, St. Paul

Morishima, M. (1982) _Why has Japan 'succeeded'?_, Cambridge University Press, Cambridge

Myers, P. (1974) So Goes Vermont, Conservation Foundation, Washington D.C.

National Agricultural Land Study (1981) Final Report, Washington D.C.

National Land Agency (1977) Japan's Metropolitan Policy, Tokyo

National Land Agency (1978) National Land Use Plan (National Plan), Tokyo

National Land Agency (1979) Sanzenso - the Third Comprehensive National Development Plan, Tokyo

National Land Agency (1980) Tsukuba - Academic New Town, Tokyo

National Land Agency (1981) The Outline of Annual Report on the National Land Use, Tokyo

National Land Agency (1982) White Paper on National Land Use, Tokyo

Nature Conservancy Council (1977) A Nature Conservation Review: Towards Implementation: A consultative paper, London

Nove, A. (1977) The Soviet Economic System, George Allen and Unwin, London

Nuti, D.M. (1982) 'The Polish Crisis: Economic Factors and Constraints' in J. Drewnowski (ed.), Crisis in the East European Economy, Croom Helm, London, pp. 18-64

Ofer, G. (1977) 'Economizing on urbanisation in socialist countries: historical necessity or socialist strategy?' in A.A. Brown and E. Neuberger (eds.), International Migration: A Comparative Perspective, Academic Press, New York, pp. 277-304

Ogura, T. (1979) Can Japanese Agriculture Survive?, Agriculture Policy Research Center, Tokyo

Okolow, C. (1980) 'National Parks in Poland', Nature and National Parks, 18, 4-6

Ophuls, W. (1977) Ecology and the Politics of Scarcity, W.H. Freeman, San Francisco

Oriental Economist (1981) Japan Economic Yearbook 1981/2, Tokyo

Ostrowski, W. (1966) 'History of Urban Development and Planning' in J.C. Fisher (ed.), City and Regional Planning in Poland, Cornell University Press, Ithaca, pp. 9-55

Pallot, J. and Shaw, D.J.B. (1981) Planning in the Soviet Union, Croom Helm, London

Pearson Commission (1969) Partners in development: report of the Commission on International Development, Pall Mall, London

Perrott, R. (1968) The Aristocrats, Wiedenfeld and Nicolson, London

Phillips, B. (1982) 'Argument delays land use paper', The Times, London, 19 July, 2.

Popper, F.J. (1981) The Politics of Land-Use Reform, University of Wisconsin Press, Madison

Prezydium Rady Narodowej Miasta Stolecznego Warszawy (1965) Plan Generalny Warszawy, Warszawa

Prime Minister's Office, Statistics Bureau (1981) Japan Statistical Yearbook 1981, Tokyo

Prince, H. and Lowenthal, D. (1965) 'English Landscape Tastes', Geographical Review, 55, 186-222

Pynoos, J., Schafer, R. and Hartman, C.W. (1980) Housing Urban America, 2nd edn., Aldine, New York

Ratcliffe, J. (1981) An Introduction to Town and Country Planning, 2nd edn., Hutchinson, London

Reiner, T.A. and Wilson, R.H. (1979) 'Planning and Decision-making in the Soviet City: Rent, Land and Urban Form' in R.A. French and F.E.I. Hamilton (eds.), The Socialist City, Wiley, Chichester, pp. 49-72

Rostow, W.W. (1975) 'Kondratieff, Schumpeter, and Kuznets: trend periods revisited', Journal of Economic History, 35, 719-753

Royal Society (1980) Environmental Aspects of Increased Coal Usage in the United Kingdom, London

Roweis,   S. and Scott,   A.   (1978) 'The Urban Land
Question' in K. Cox (ed.), Urbanization and Conflict
in Market Societies, Methuen, Chicago, pp. 38-75

Rural Development Planning Commission (1981) Rural
Planning and Development in Japan, Tokyo

Sakiyama,   T.   (1979) 'Policies on pollution,
aquaculture and coastal management in Japan', Marine
Policy, 3, 20-24

Sampson,   S.L. (1979) 'Urbanization - Planned and
Unplanned: a Case Study of Brasov, Romania' in R.A.
French and F.E.I. Hamilton (eds.), The Socialist
City, Wiley, Chichester, pp. 507-524

Sargent,   J.   (1980) 'Industrial Location in Japan
since 1945', GeoJournal, 4, 205-214

Scott   (1942) Report of the Committee on Land
Utilisation in Rural Areas, Cmd 6378, HMSO, London

Scott,   A.J.   (1980) The Urban Land Nexus and the
State, Pion, London

Scottish   Development   Department (1977) National
Planning Guidelines, Edinburgh

Scottish   Development   Department (1981) National
Planning Guidelines (Revised), Edinburgh

Self, P. (1975) Econocrats and the Policy Process:
the Politics and Philosophy of Cost-Benefit
Analysis, Macmillan, London

Sharpe, J. (1975) 'Innovation and change in British
land-use planning' in J.   Hayward and M. Watson
(eds.), Planning, Politics and Public Policy,
Cambridge University Press, Cambridge, pp. 316-357

Shaw, D.J.B. (1981) 'Problems of land use planning
in the USSR', Soviet Geography, 22, 293-305

Shoard,   M. (1981) The Theft of the Countryside,
Morris Temple Smith, London

Shulstad, R.H., Herrington, B.E., May, R.D. and
Rutledge,   E.M.   (1980) 'Estimating a potential
cropland supply function for the Mississippi delta
region', Land Economics, 56, 457-464

Simmie,   J.M.   (1974) Citizens in Conflict: the

sociology of town planning, Hutchinson, London

Simon, J.C. (1981) The Ultimate Resource, Princeton University Press, Princeton

Siuty, J., ed. (1978) Ochrona i rekultywacja gleb, Panstwowe Wydawnictwo Rolnicze i Lesne, Warszawa

Smith, J.S. (1981) 'Land transfers from farming in Grampian and Highland, 1969-1980', Scottish Geographical Magazine, 97, 169-174

Soil Conservation Society of America (1977) Land Use: Tough choices in today's world, Iowa

Strathclyde Regional Council (1979a) Strathclyde Regional Planning Survey 1979, Glasgow

Strathclyde Regional Council (1979b) Structure Plan, Written Statement, Glasgow

Strong, A.L. (1981) 'Land as a Public Good' in J.I. de Neufville (ed.), The Land Use Policy Debate in the United States, Plenum Press, New York, pp. 217-232

Susskind, L. (1981) 'Citizen Participation and Consensus Building in Land Use Planning' in J.I. de Neufville (ed.), The Land Use Policy Debate in the United States, Plenum Press, New York, pp. 183-198

Sutcliffe, A. (1981) Towards the Planned City: Germany, Britain, the United States and France 1780-1914, Basil Blackwell, Oxford

Tajima, K. (1978) 'New lanterns for old', Built Environment Quarterly, 4, 31-35

The Council of State Governments (1982) Forest Resource Management: Meeting the Challenge in the States, Lexington

The Times (1983), 28 January, p. 10

Tokyo Metropolitan Government (1976) Land in Tokyo Metropolis in 1974: Aspect and Program, Municipal Library No. 12, Tokyo

Tokyo Metropolitan Government (1982) Tokyo tomorrow, Municipal Library No. 17, Tokyo

Tokyo Metropolitan Research Center for Environmental

Protection (1979) <u>Environmental Data Book for Tokyo</u>,
Tokyo

Tunbridge,   J.E.   (1981)   'Conservation   trusts   as
geographic   agents: their   impact   upon   landscape,
townscape and land use', <u>Transactions, Institute of
British Geographers</u>, <u>6</u>, 104-125

United   Nations,   Department of Economic and Social
Affairs   (1973) <u>Urban   Land   Policies   and   Land-Use
Control Policies</u>, vol.   III <u>Western Europe</u>, New York

United   Nations,   Department of Economic and Social
Affairs   (1975) <u>Urban   Land   Policies   and   Land-Use
Control Policies</u>, vol.   IV, New York

United Nations (1983) <u>Demographic Yearbook 1981</u>, New
York

United States Department of Commerce,   Bureau of the
Census (1980) <u>Taxes and Intergovernmental Revenue of
Counties,   Municipalities and Townships: 1978-79</u>,
Washington D.C.

Uthwatt   (1942) <u>Report   of   the Expert Committee on
Compensation and Betterment</u>, Cmd 6386, HMSO, London

Walker,   R.A.   and   Neiman,   M.K.   (1981)   'Quiet
revolution   for   who?', <u>Annals</u>,   Association   of
American Geographers, <u>71</u>, 67-83

Williams, N. (1981) <u>American Planning Law - Land Use
and the Public Law</u>, 5 vols., Callaghan, Chicago

World   Bank   (1979,   1980,   1981) <u>World Development
Report 1979</u>, <u>1980</u>, <u>1981</u>, Washington D.C.

Yanaga, C. (1968) <u>Big Business in Japanese Politics</u>,
Yale University Press, New Haven

Zielinski,   J.G.   (1968) <u>On the Theory of Socialist
Planning</u>, Oxford University Press, Ibadan

Ziemetz,   K.A.,   Dillon,   F.,   Hardy, E.E. and Otte,
R.C.   (1976) <u>Dynamics   of   Land Use in Fast Growth
Areas</u>,   United States Department   of   Agriculture,
Economic   Research   Service,   Agricultural Economic
Report No.   325, Washington D.C.

INDEX

Printed and bound by CPI Group (UK) Ltd, Croydon, CR0 4YY

22/10/2024

01777615-0004